Evolutionary ecology of marsupials

T0269316

MONOGRAPHS ON MARSUPIAL BIOLOGY

Evolutionary ecology
of marsupials

ANTHONY K. LEE

Associate Professor, Department of Zoology, Monash University, Clayton, Victoria, Australia

ANDREW COCKBURN

Lecturer, Department of Zoology,
Australian National University, Canberra, Australia

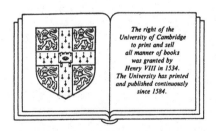

The right of the
University of Cambridge
to print and sell
all manner of books
was granted by
Henry VIII in 1534.
The University has printed
and published continuously
since 1584.

CAMBRIDGE UNIVERSITY PRESS

Cambridge
New York   New Rochelle   Melbourne   Sydney

CAMBRIDGE UNIVERSITY PRESS
Cambridge, New York, Melbourne, Madrid, Cape Town, Singapore, São Paulo

Cambridge University Press
The Edinburgh Building, Cambridge CB2 8RU, UK

Published in the United States of America by Cambridge University Press, New York

www.cambridge.org
Information on this title: www.cambridge.org/9780521252928

First published 1985
Reprinted 1987
This digitally printed version 2008

A catalogue record for this publication is available from the British Library

Library of Congress Catalogue Card Number: 84-9440

ISBN 978-0-521-25292-8 hardback
ISBN 978-0-521-05412-6 paperback

Contents

Preface

Our purpose in this text is to place the natural history of marsupials within an evolutionary framework. We feel that a synthesis of this sort is warranted for four reasons. First, we freely admit that we are biased towards the view that the questions of greatest utility in ecological theory are those framed in an evolutionary context, and that we are thus concerned with the ways in which individuals in different environments enhance their survival and reproduction. This perspective has been applied to marsupial reproduction only occasionally. Second, most of the literature on marsupial ecology continues to be descriptive, its authors being motivated by the dearth of natural history data for most species. While this descriptive information is desperately needed, the absence of a comparative framework has led to the use of disparate and incomparable methods and this has hindered synthesis. Third, where evolutionary theory has been applied to marsupials, it has often been done unwisely, using assumptions now discarded by evolutionary biologists. Last, we believe that the unique mode of reproduction in marsupials affords an opportunity to examine a number of problems of general relevance. We show that behavioural and ecological convergence between marsupials and eutherians has been as frequent as morphological convergence. But because marsupials have a short gestation and a period of obligatory attachment to the nipple after birth, a greater period of the life history is accessible for observation and manipulation than is the case for eutherians. Matrilineal and sibling relationships can be established with unusual confidence by marking and sexing pouch young, without any risk of brood abandonment resulting from nest disturbance. Such an understanding of kinship can be used to shed light on evolutionary questions relevant to mammals generally.

While our emphasis is thus different from the traditional thrust of natural history studies on marsupials, we acknowledge our indebtedness to the naturalists who gathered the data on which we depend. Our task in collating these data was greatly facilitated by the timely publication of important reviews by Eleanor Russell (1982a) and John Eisenberg (1981). Although we occasionally disagree with the thrust of their arguments, both contributions stimulated much of our analysis and discussion.

We have used the nomenclature of Kirsch & Calaby (1977) except where it has been revised by Archer (1982) for the Dasyuridae and by McKay (1982) in the use of *Petauroides* for *Schoinobates*. Common names are used only where these are in general use.

One of us (A.C.) was supported during the long task of writing by a CSIRO Postdoctoral Fellowship at the Museum of Vertebrate Zoology, University of California at Berkeley; a Monash Postdoctoral Research Fellowship at the Department of Zoology at Monash University; and a Queen Elizabeth II Fellowship at the CSIRO Division of Wildlife and Rangelands Research and the Department of Environmental Biology, Research School of Biological Sciences, Australian National University.

Our colleagues at these institutions were a constant source of data, references and stimulating ideas. At Monash, many of our ideas were first floated at the Mammal Discussion Group. Henri Frey, Mike Fleming, Stephen Henry, Brian Malone, Roger Martin, Peter Mitchell, Ken Nagy, John Nelson, Anne Opie, Lester Pahl, Greg Parry, Marilyn Renfree, Gordon Sanson, Michelle Scott, Roger Short, Andrew Smith, Mike Stoddart and Vivienne Turner were especially important contributors, either of criticism or of unpublished data (or both). Special gratitude is owed to Dick Braithwaite, who single-handedly interested both of us in the questions which now form the central theme of our research. The academic environment at Berkeley is justifiably famous, as is its library, and the countless discussion groups and seminars were particularly stimulating. Pedro Alberch, Herbert Baker, Irene Baker, Rob Colwell, Ed Heske, Suzanne Koptur, Bill Lidicker, Katie Milton, Oliver Pearson, Paul Sherman and Dave Wake all introduced us to exciting new areas of interest.

Pierre Charles-Dominique and Eleanor Russell kindly allowed us access to their unpublished papers. We accept responsibility for any errors in translation from French or Spanish, but thank Henri Frey and Paula Henriksen for help in this regard. Our final special thanks are to Hugh Tyndale-Biscoe for his constructive criticism of the entire manuscript, and for his liberal interpretation of the word deadline.

October 1983   A.K.L. & A.C.

1

Introduction

Evolutionary ecology is concerned with the ways in which organisms living in different environments enhance their survival and reproduction. In this text we consider the evolutionary ecology of the marsupials, a group of mammals which have fascinated biologists since the discovery of their unique mode of reproduction. In doing this we do not offer an exhaustive review of all aspects of ecology. Rather, we focus on instances where marsupials elucidate problems in evolutionary ecology and vice versa. Most of our endeavour must be viewed as an attempt to place descriptive data in a standard theoretical framework which we hope leads naturally to the generation of hypotheses which examine the resilience of the framework and provide new directions in marsupial research.

Origins of theoretical ecology

Ernest Haeckel (1866) coined the term 'Oekologie' to embrace animal/environment relationships while discussing animal morphology in the light of Charles Darwin's new theory of evolution by natural selection. Despite the early importance of evolutionary theory in distinguishing ecology as a science, a theoretical basis for much of ecological research has been lacking. McIntosh (1980) points out that there were only two references to theory and one to hypothesis in the pre-1950 cumulative indices of the major American ecological journals *Ecology* and *Ecological Monographs*, but that since that time a theoretical literature has burgeoned.

Recent historians of science have commented that this literature gives two distinct views of the organisation of ecosystems, with different historical bases (Ghiselin, 1974; MacFayden, 1975; F. E. Smith, 1975; Harper, 1977; McIntosh, 1980; Simberloff, 1980). These are:

1

1 *Systems ecology*, which sees ecosystems as developing evolutionary entities which guide the evolution of species but which are to some extent independent of their constituent species. This view has a recent antecedent in F. E. Clements' (1905) now-discredited 'organismic' view of the community, and some authors have traced its origins to Platonic idealism and Aristotelian essentialism, which formed the predominant western pre-Darwinian view of nature (e.g. Ghiselin, 1974; Simberloff, 1980).

2 *Evolutionary ecology*, which treats the ecosystem as the sum of its parts (species), with understanding of the whole system deriving from studies of species populations and their characteristics as products of natural selection. This philosophy is clearly post-Darwinian, and much of its history is eloquently reviewed by Hutchinson (1978).

While we agree with Levins & Lewontin (1980) that properties of ecosystems and communities are objects of interest in their own right, the ability to synthesise these two views of ecosystems has remained elusive (see e.g. Wilson, 1980). In this book we adopt the second view above, and argue that evolutionary theory represents the chief unifying perspective through which natural history data can be usefully interpreted, and around which predictive hypotheses can be formulated (see also Southwood, 1980).

Although Darwin's ideas were widely understood by the end of the nineteenth century, the reinterpretation of the different subdisciplines of biology in an evolutionary framework has proceeded at different rates in different fields. Historians of science attribute particular importance to the synthesis of the concepts of natural selection and Mendelian genetics in the 1920s and 1930s by authors such as R. A. Fisher, S. Wright and J. B. S. Haldane (see Mayr & Provine, 1980, for a full discussion). It was not immediately obvious how population processes and species interactions related to an evolutionary theory that was based on the reproductive success and survival of individuals, despite explicit considerations of ecological questions by Darwin and Fisher. This led to the persistence of a theoretical vacuum for the interpretation of life histories and behaviour, and to the occasional misapplication of evolutionary theory, especially through an undue emphasis on group selection (e.g. Wynne-Edwards, 1962). New ways of resolving these difficulties were provided by, among others, D. Lack (1947, 1954, 1968), G. E. Hutchinson (1957, 1959) and R. H. MacArthur (1972), who examined the causes and consequences of variation in the life history of organisms and the responses of individuals to competition and other species interactions, and by G. C. Williams (1966), who critically reviewed the conditions for group and individual selection.

Much of the work after this time was couched in simple mathematical terms (for a synopsis see Roughgarden, 1979) and provided a theoretical framework which has led to an expansion of the application of mathematical solutions to ecological problems. It also contributed to the intense interest now shown in evolutionary ecology. Unfortunately, the mathematical content of this later work either discouraged the natural historians who had generated the descriptive data which the theories attempt to place within a coherent framework, or occasionally led to uncritical acceptance of mathematical predictions of dubious ecological relevance. This has led to an increasing gap between theorists and empiricists. As one of the founders of evolutionary ecology, G. E. Hutchinson (1975, p. 516) has commented:

> Many ecologists of the modern generation have great ability to handle the mathematical basis of the subject. Modern biological education, however, may let us down as ecologists if it does not insist... that a wide and quite deep understanding of organisms, past and present, is as basic a requirement as anything else in ecological education. It may be best self-taught, but how often is this difficult process made harder by a misplaced emphasis on a quite specious modernity.

The ecology of marsupials

Only a few ecological studies of marsupials have had some theoretical content. Main, Shields & Waring (1959) were able to identify three motivations for early ecological research: economic pressure for control of pest species (including grazing kangaroos and feral marsupials in New Zealand), conservation of well-known but endangered species and some academic questions, particularly those related to how populations behave. The most influential academic study of this sort had its genesis during the 1950s, and began as an investigation of population 'crashes' in the quokka (*Setonix brachyurus*) on Rottnest Island (see Main *et al.*, 1959). This research laid the foundations for a number of investigations of the ecological strategies exhibited by macropods (e.g. by Newsome, 1965; Main, 1971; Main & Bakker, 1981), and the Macropodidae continues to be the best understood of the various families of marsupial.

A convenient starting-point from which to examine the impact of evolutionary theory on more recent studies is the year 1968, following the

publication of the classic texts *Adaptation and Natural Selection*, by
G. C. Williams (1966), and *The Theory of Island Biogeography*, by R. H.
MacArthur & E. O. Wilson (1967), which heralded a new maturity in the
application of evolutionary theory to ecological questions. It is clear from
Fig. 1.1 that the amount of research which focused on management and
the clarification of distribution fluctuated irregularly during the period
1968–82. By contrast, there was a proliferation of studies with a more
purely academic orientation in the latter years, though only a few of these
discuss data in an evolutionary context. Where this is the case, however,
most invoke evolutionary explanations *a posteriori*, rather than seek to test
hypotheses pertinent to evolutionary theory.

Synopsis of argument

The explosion of interest in evolutionary ecology has very recently
influenced the course of marsupial research. This book consists of six
loosely-connected essays which attempt to integrate these fields. The
arguments we put forward are summarised below.

Throughout the text we stress the importance of the quality of food and

Fig. 1.1. Trends in Australian research on marsupial ecology during
the period 1968–82. Data obtained from Calaby, J. H., Current
Literature on Marsupials, an occasional series in *Australian
Mammalogy* and *Bulletin of the Australian Mammal Society*.

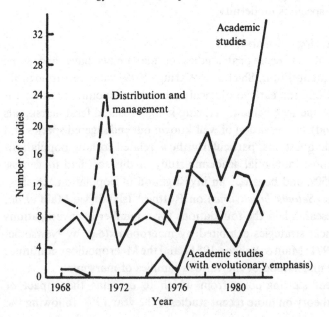

of its dispersion in space and time for the evolution of marsupial life histories and social behaviour. In Chapter 2 we summarise the range of foods exploited by marsupials and the problems associated with the procurement and processing of different foodstuffs. The influence which body size has on the efficiency of procurement and processing is stressed, and the range of body sizes of marsupials with different feeding strategies is identified.

Because fitness will ultimately be expressed in terms of changes in survival and/or reproduction, variations in these parameters (life history traits) have attracted considerable attention in evolutionary theory. The chief difference between eutherians and marsupials is the timing of reproduction and development, with marsupials investing comparatively little in gestation. We promote the view that organisms must be viewed as a combination of recent adaptation and old constraints, and that constraints may restrict diversification of higher taxa.

In Chapter 3 we contrast the eutherian and marsupial radiations, and conclude that the marsupials may be distinguished by a slower 'pace of life' and conservative morphological and behavioural variation. Although causes for marsupial conservatism have been suggested, none are strongly supported by available data. We develop a case which suggests that specialisation for early extrauterine life may have restricted the capacity of marsupials for paedomorphic change. Marsupials should thus be viewed as specialised, rather than primitive.

In Chapters 4 and 5 we examine the adaptation of life history traits within the constraints identified in Chapter 3. In Chapter 4 we examine the predominantly carnivorous and omnivorous polyprotodont marsupials. Data are available for three families, the Peramelidae, the Dasyuridae and the Didelphidae. All three occupy many habitats across a broad geographic range. The Peramelidae show little variation in life history traits, but can be distinguished from other marsupials by their rapid reproductive rates. By contrast, the Dasyuridae exhibit a variety of life history strategies, and these appear to be explicable in terms of the effect of body size on reproductive rate and the predictability and seasonal availability of food. The bandicoots (peramelids) mainly show facultative changes, while the life histories of dasyurids are more rigidly controlled. Studies of the Didelphidae have been restricted to tropical species, and so we are unable to place them within this perspective. However, recent data suggest that quality of food determines their reproductive rate.

The herbivorous, nectarivorous and plant exudate-feeding diprotodont marsupials are considered in Chapter 5. They show considerable diversity in both habitats occupied and life history traits but are less fecund than most polyprotodont marsupials. Once again the variety of life history strategies is largely explicable in terms of the constraints of body size, differences in the quality of food and the seasonal availability of food. There is less variation in litter size than in the Dasyuridae and litters are generally smaller. Instead variation in the interval between litters assumes greater importance in determining fecundity. In the Tarsipedidae, Burramyidae and Macropodidae, this interval may be reduced by over-lapping consecutive litters. This overlap is considered to be one of the ecological values of embryonic diapause, which is found only in these taxa among marsupials. Social organisation is documented for a number of these diprotodont marsupials and is shown to be more diverse than previously recognised. Variation in social organisation is related to the quality and dispersion of food.

Marsupials may be suitable for further investigation of variation in life history and social organisation as the young can be marked and sexed during the period of obligatory attachment to the teat. This means that genetic relationships and the natal site of weaned young can be established with unusual confidence. In Chapter 6 we illustrate and expand upon the nature of this advantage by discussing our ongoing research on *Antechinus*. Australian species of this genus have the simplest life history known among higher vertebrates, as synchronous, seasonal breeding and complete post-mating male mortality mean that there is never more than one cohort of males and generally not more than three cohorts of females (often less) at any one time. This simplicity and synchrony, coupled with the ability to mark and sex young, has provided the opportunity for investigating topics not normally tractable in empirical studies of mammals. We discuss the relationship between reproductive effort and mortality, the evolution of geographic variation in brood size and the evolution of extraordinary sex ratios.

In Chapter 7 we go on to consider the impact of coevolution on the structure of communities of species. Mutualisms involving marsupials which require further investigation include plant/pollinator, fruit/frugivore, grass/grazer and fungus/fungivore interactions. The evidence for coevolu-tion in these categories is reviewed, and the need for quantitative studies of the impact of marsupials on plant reproductive success is stressed. Recent attempts to invoke competition between species as a determinant of the structure of marsupial communities and the morphology and behaviour of constituent species are reviewed, and placed in the perspective of the ongoing controversy about the role competition plays in structuring communities. The need for experimental studies is stressed.

We conclude by suggesting a number of profitable avenues of inquiry which would fill in gaps in the knowledge discussed in this book or test hypotheses that we feel are worth further investigation.

2

*Marsupials and their resources**

It is presumed that natural selection operates to ensure production of the maximum number of young who survive to maturity, summed over a lifetime. The ability of individuals to maximise their production of young is dependent upon the rate at which they can procure resources and the way in which they manipulate those resources to produce offspring. In

Table 2.1. *Terms used to describe activity substrates of marsupials*

Fossorial	Marsupials that conduct most of their life-supporting activities underground, including foraging, nesting, mating and rearing of young. They may venture to the soil surface and may even feed there
Semifossorial	Marsupials that use burrows as refuges or obtain most of their food by digging but which move freely on the soil surface
Semiaquatic	Marsupials that carry out life-supporting activities in water but spend at least a portion of each 24-hour period out of water
Terrestrial	Marsupials that forage primarily on the soil surface and have limited ability to burrow
Scansorial	Marsupials that show considerable adaptation for climbing but which spend approximately equal time on the ground and in trees
Arboreal	Marsupials that spend most of their lives in trees

After Eisenberg (1981).

* Much of the information in this chapter is drawn from an unpublished manuscript by A. Smith, G. D. Sanson & A. K. Lee.

mammals the essential resources are food and mates, and to a lesser extent nest sites. The rate at which animals can procure resources depends upon the dispersion of those resources in time and space, and the rate at which they need to harvest food depends upon its quality and their requirements.

Our objectives in this chapter are to describe the resources (particularly food) used by marsupials and to consider the quality and dispersion of those resources. We introduce the major taxa of marsupials and describe their habitats (activity substrates) (Table 2.1) and activity rhythms (Table 2.2). Eisenberg (1981) classified mammals according to their dietary specialisations (Table 2.3) and substrate use (Table 2.1), and we have adapted his terminology for these descriptions. Certain of Eisenberg's categories, such as the substrates aquatic and volant, and the feeding

Table 2.2. *Terms used to describe activity rhythms of marsupials*

Nocturnal	Activity principally confined to the hours of darkness
Crepuscular	Activity peaking at dawn and dusk
Even diel cycle	Activity distributed throughout day and night
Diurnal	Activity principally confined to daylight hours

Table 2.3. *Feeding categories of marsupials*

Feeding category	Principal food types (other food)
Carnivore	Terrestrial vertebrates
Insectivore/omnivore	Arthropods (molluscs, earthworms, flowers)
Myrmecophage	Social insects, particularly ants and termites
Exudate feeder/insectivore	Plant exudates such as sap, gum and resins (insects)
Fungivore/omnivore	Hypogeous fungal sporocarps (insects, gum)
Frugivore/omnivore	Fruit (insects, exudates)
Frugivore/granivore	Fruit (seeds)
Nectarivore	Nectar and pollen
Browsing herbivore	Leaves, buds and stems of dictotyledonous plants
Grazing herbivore	Grasses

After Eisenberg (1981).

categories piscivore and aerial insectivore, are omitted because no marsupial falls within them; others have been adjusted to describe the marsupial mode (e.g. fungivore/omnivore) more clearly. As Eisenberg warns, there are many mammals which are difficult to place precisely in one or other of these categories, and many conform only if we think in terms of modal tendencies during the annual cycle.

Marsupial taxa
Didelphidae
There are approximately 70 species of didelphids; they range in size from about 0.01 kg in *Thylamys* and *Marmosa* to about 2 kg in *Didelphis* (see Fig. 2.1 for body length distribution). They are distributed widely in the Americas but attain maximum diversity in tropical South America. They inhabit all of the biomes of South America from desert and alpine habitats to wet tropical forests. McNab (1982) has commented that few species have invaded temperate environments in South America despite the long availability of such habitat. Some forms, for example *Monodelphis* and *Didelphis*, have a generalised morphology which is thought to resemble the ancestral proto-mammalian condition. Morphological adaptations are pronounced in the semiaquatic *Chironectes*, which has webbed hind-feet, in *Lutreolina*, which resembles a mustelid, and in the Caluromyinae (*Caluromys*, *Caluromysiops* and *Glironia*), which are highly arboreal. Other species exhibit different levels of arboreality, which are reflected in characters such as tail-length and prehensility (Eisenberg & Wilson, 1981). Arboreal forms resemble some of the nocturnal prosimians. *Metachirus* and *Monodelphis* are the most terrestrial genera among the didelphids. All species are predominantly nocturnal.

The diets of most of the species are not really known, but it seems safe to conclude that most are insectivore/omnivores and feed opportunistically. *Monodelphis* is highly insectivorous, *Lutreolina* may be more carnivorous than other genera, and *Chironectes* eats a variety of aquatic invertebrates and vertebrates (Hunsaker, 1977). Ground-dwelling genera eat carrion and some live vertebrate prey. Arboreal species are more frugivorous than terrestrial species, and the Caluromyinae appear to eat fruit from many taxa (Charles-Dominique, 1983). Arboreal and scansorial species supplement their diet with nectar (families known to be visited include Bombacaceae, Euphorbiaceae, Balanophoraceae, Mimosaceae and Caesalpinaceae), and *Caluromys* has been observed to eat the gum of *Fagara rhoifolia* (Rutaceae) during periods of fruit shortage (Charles-Dominique et al., 1981).

Fig. 2.1. Distribution of body lengths (snout-vent; mm) of some marsupial families. (*a*) Dasyuridae (stippled), Thylacinidae (open); (*b*) Peramelidae (stippled), Thylacomyidae (open); (*c*) Didelphidae (coarse stippled), Microbiotheriidae (fine stippled), Caenolestidae (open). For data see Appendix 1.

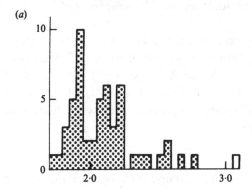

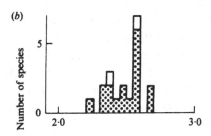

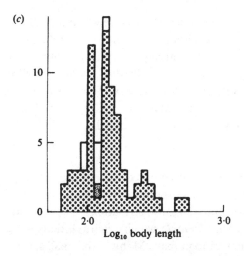

Little is known of feeding behaviour, though *Marmosa cinerea* is highly mobile and agile during prey capture, while *Caluromys* moves more slowly, with the ears and head moving continuously, suggesting that hearing plays a major role in prey capture (Charles-Dominique *et al.*, 1981). Olfaction is clearly important in prey location in *Monodelphis* and *Didelphis* (Streilein, 1982a).

A number of studies in tropical forests suggest that *Didelphis marsupialis* is partially nomadic, moving between nests, which are often found in tree cavities (Davis, 1945b; Miles, de Souza & Povoa, 1981). An account of nest sites of other species is provided by Miles *et al.* (1981), but the extent of nomadism is not known. *Caluromys* appears to be more sedentary than *Didelphis* (Atramentowicz, 1982).

Caenolestidae

There are seven species in this family; each weighs about 0.04 kg (see Fig. 2.1 for body length distribution). *Caenolestes* and *Lestoros* occur along the Andes Cordillera from Venezuela to Peru, while *Rhyncholestes* is found in the Valdivian rainforest of Southern Chile. All species prefer densely-vegetated, cold and wet forests at elevations ranging from sea level to more than 4200 m. They are common on moist, moss-covered slopes and ledges (Marshall, 1982). Caenolestids are shrew-like, terrestrial and crepuscular or nocturnal (Kirsch & Waller, 1979), and are probably insectivore/omnivores. No comprehensive ecological or behavioural study has been undertaken on any species.

Microbiotheriidae

The one extant species, *Dromiciops australis*, weighs about 0.04 kg (see Fig. 2.1 for body length) and occurs in the Valdivian rainforest of Chile. *Dromiciops* is nocturnal and feeds largely on the larvae and pupae of a variety of insects (see e.g. Fischer, 1978; Marshall, 1978a). Its foot morphology suggests arboreality (Szalay, 1982). In view of the controversial taxonomic position of this species (see e.g. Sharman, 1982; Szalay, 1982), a study of the ecology and behaviour is urgently needed.

Dasyuridae

The 52 extant species in this family embrace the body-weight range 0.003–8 kg. Seventy-five per cent of these species have a body weight of less than 0.1 kg (see Fig. 2.1 for body length distribution). They include the smallest marsupials, in the genus *Planigale*, which weigh as little as 3–4 g as adults and are among the smallest mammals. Many of the small species resemble shrews (Insectivora) and tree shrews (Scandentia), while the largest dasyurids have counterparts among the mustelids and small viverrids (Carnivora). There is surprisingly little variation in body form.

Dasyurids are found in most terrestrial habitats in Australia and New Guinea, which, in Australia, encompass the entire latitudinal and altitudinal range. They are particularly successful in arid Australia, where they are represented by 19 species, or 47% of Australian dasyurids, a higher proportion than any other mammalian family (Morton, 1982). They are also successful in forests, where *Antechinus* spp. are often the most abundant terrestrial mammals.

Dasyurids are terrestrial or scansorial, and principally nocturnal or crepuscular. A few species such as *Antechinus swainsonii* and *A. minimus* feed throughout the diel cycle. Hall (1980a) suggested that the different activity patterns of *A. stuartii* (see Plate 2.1), which is primarily nocturnal, and *A. swainsonii* are related to the accessibility of their prey. *A. stuartii* is thought to feed upon prey which are active on the soil surface at night and inaccessible during the day, whereas *A. swainsonii* is able to locate and feed upon concealed prey. The very small *Planigale* and *Ningaui* spp., which have high energy demands, may also be active throughout the diel cycle, interspersing bouts of activity with short periods of rest and torpor (Morton & Lee, 1978).

There is comprehensive information on the diets of only a few dasyurids.

Plate 2.1. *Antechinus stuartii*, a nocturnal insectivorous dasyurid. (Jill Poynton.)

All the species appear to be opportunists, taking a broad range of dietary items in roughly the proportions available to them. Arthropods, particularly insects, are common to all the diets, but decrease in their proportional contribution with an increase in species size of the dasyurid. Small dasyurids such as *A. stuartii* and *Sminthopsis crassicaudata* feed almost exclusively on arthropods, vertebrate remains being only occasionally found in their faeces (Morton, 1978a; Hall, 1980b; Statham, 1982). The larger dasyurids such as *Dasyurus viverrinus* feed on vertebrates as well as arthropods, and the largest dasyurid (*Sarcophilus harrisii*) subsists primarily upon mammalian carrion, particularly of the wallaby *Macropus rufogriseus*, the wombat *Vombatus ursinus* and sheep (Guiler, 1970a). The larger dasyurids also tend to take fewer kinds of items, and especially favour those that are locally common. For example Blackall (1980) found that the larvae of moths and adult beetles which were particularly common in grass tussocks were also the most common dietary items of *D. viverrinus*. The moth larvae emerge to forage on pasture at night and can be harvested in substantial numbers. Evidence of birds and mammals was found in only 17% of the faeces of this species, whereas insects, especially moth larvae and beetles, were found in 97% of the scats.

Plant material occurs in the diets of all the species. The fruit of *Rubus* spp. (Rubaceae) eaten by *A. swainsonii* (A. K. Lee, personal observation), and the flowers of *Acrotriche aggregata* (Epacridaceae) eaten by *Antechinus stuartii* (Statham, 1982), appear to be taken selectively. Other plant material may be derived from the gut contents of digested pasture-feeding larvae such as those eaten by *D. viverrinus*, or may be eaten along with these food items. Blackall (1980) observed that moth larvae often cling tenaciously to grass stems when disturbed and are consumed along with the grass stems. Up to 60% of the volume of the scats of *D. viverrinus* consisted of grass, leaves, seeds and other plant components, but it is doubtful whether this plant material contributes at all to their nutrition.

Little is known of the modes of prey hunting or capture in dasyurids. *A. swainsonii* forages among forest litter, sometimes digging to obtain prey. *A. stuartii* appears to hunt for items at the soil surface, darting from item to item, and also searches beneath the loose bark on trees. The prey-capturing skills of the larger dasyurids differ markedly. *S. harrisii* is slow and clumsy when killing rats, which is consistent with its dependence on vertebrate carrion, whereas *D. viverrinus* is quick and adept at killing rats, and at climbing, and probably hunts for small birds and mammals (Buchmann & Guiler, 1977).

Dasyurids use a variety of nest sites including tree hollows (*A. stuartii*;

Dickman, 1982a), burrows or hollows in rotting logs (*A. swainsonii*; Hall & Lee, 1982), rockpiles and hollow logs (*D. viverrinus*; J. Godsell, personal communication), and rock crevices (*Parantechinus bilarni*; Begg, 1981a).

Thylacinidae

Thylacinus cynocephalus was the largest of the Recent marsupial carnivores, weighing between 15 and 35 kg. Superficially, it resembled a small dog, standing about 60 cm at the shoulder. Found as a fossil at widely scattered localities in Australia and New Guinea, it was confined historically to Tasmania, where it now appears to be extinct (R. J. Smith, 1981).

There is little information on diet. It has been suggested that thylacines hunted alone or in pairs, and that before the arrival of European settlers in Tasmania they fed upon wallabies, possums, bandicoots, rodents and birds (R. J. Smith, 1981). It is also thought the thylacines caught prey by stealth rather than by chase.

At the time of European settlement, *T. cynocephalus* was widespread in Tasmania and was particularly common where pasture adjoined dense forest. It nested during the day in dense forest on hilly terrain and emerged at night to feed in grasslands and woodlands. Its decline in Tasmania primarily resulted from persecution as a consequence of its reputation for killing sheep (R. J. Smith, 1981).

Notoryctidae

Little is known about *Notoryctes typhlops*, the sole species in this family, and the only fossorial marsupial. It resembles the golden moles of Africa (Chrysochloridae) and true moles (Talpidae) in appearance. The nails of the second and third toes of the front feet are greatly expanded and serve as shovels. The snout is covered by a horny shield, the only evidence of the external ear is a small hole and the eyes are not visible. While it undoubtedly spends most of its time beneath the soil surface in the sandy deserts of central Australia, specimens are occasionally captured at night at the surface (Corbett, 1975). Analysis of the gut contents of two specimens suggests that the species feeds on ants (*Rhytidoponera* sp.), chrysomelid beetles and the subterranean larvae of sawflies (Corbett, 1975). Scarabaeid beetle larvae may also be an important food resource.

Myrmecobiidae

Myrmecobius fasciatus, is a myrmecophage, feeding primarily on termites extracted from shallow subterranean galleries. The proportion of different termite species in the diet corresponds with their natural abundance

(Calaby, 1960). Calaby found that ants, mostly small predatory species, comprise about 15% of the diet. He considered that these species had been ingested accidentally either while they were swarming over termites in the broken galleries or when nests adjacent to the termites' galleries were accidentally broken into. Like other mammalian myrmecophages, this species has a long snout, a long protrusible tongue and small degenerate teeth. Prominent claws on the forefeet are used to excavate galleries.

Historically, this species occurred over a substantial part of southern Australia, but is now restricted to the south west of Western Australia. It is confined today to open forest and woodland where the forest floor is littered with fallen branches and hollow logs. Hollow logs are used for shelter and nest sites. It is the only diurnal marsupial.

Peramelidae

The 15 extant species of peramelids (bandicoots) occur within a geographic range encompassing Australia, New Guinea and adjacent islands. Bandicoots are small-to-medium in size (0.15–5 kg; see Fig. 2.1. for body length distribution). They are semifossorial, crepuscular or nocturnal, and bear some resemblance to small macropods. They have pointed snouts and feet with strong claws and, like macropods, have the second and third toes of the elongate hind-feet united except for the claws and the terminal portions. They are found in a variety of plant communities ranging from tropical rainforest to heathland and desert steppe.

Bandicoots are insectivore/omnivores. In his study Heinsohn (1966) found that *Perameles gunnii* and *Isoodon obesulus* (see Plate 2.2) in Tasmania fed primarily on earthworms, adult beetles, scarabaeid beetle and moth larvae, and moth pupae. Substantial quantities of the fruit of blackberry (*Rubus fruticosus*) and boxthorn (*Lycium* sp.; Solanaceae) were eaten when in season.

At a heathland site in Victoria, the diet of *I. obesulus* consisted mainly of a root-feeding Melolonthine larvae (Scarabaeidae), in autumn and winter, and ant pupae, adult scarabaeids and adult hemipterans in spring and summer (A. M. Opie, personal communication). Diptera and tene-brionid larvae, earwigs and spiders were also taken. During winter and spring, when the soil was moist, the subterranean fruiting bodies of hypogeous sporocarpic fungi (mainly *Endogone* and *Glomus*) were fre-quently eaten. Bandicoots also consume the fruit of *Cassytha* sp. (Lauraceae).

I. obesulus obtains subterranean prey by scratching conical holes in the soil up to 8 cm deep and, judging from the diet, obtains most of its food items in this way. However, the inclusion of items such as adult scarabaeid beetles and *Cassytha* fruit indicates that some food items are collected from the soil surface.

I. obesulus makes flat nests of sticks, which are sited in dense vegetation. Ride (1970) pointed out that there is no permanent entrance to the nest. The animal burrows into the nest and then conceals the entrance, the reverse procedure being adopted on departure. It may construct burrows in sandy soils during hot weather (J. Kirsch in Ride, 1970).

Thylacomyidae

The two species of thylacomyids (bilbies) are confined to the central Australian deserts. They resemble bandicoots in appearance and size (0.36–2.5 kg; see Fig. 2.1 for body length distribution), but have longer ears and very long claws on the forefeet. Bilbies are strictly nocturnal, resting during daylight hours in 1- to 2-m deep spiral burrows. These burrows may be blocked by plugs of sand when a bilby is in residence.

K. A. Johnson (1980a) analysed the faeces of *Macrotis lagotis* and found them to consist of seeds, fruit, bulbs, insects and a small amount of fungus. Approximately one-third of the faecal fragments were derived from insects (primarily ants, termites and beetles). The faecal contents revealed no

Plate 2.2. *Isoodon obesulus*, a heathland dwelling bandicoot. (Bert Lobert.)

obvious preference for particular food items, and the variation in diet appeared to reflect seasonal and local availability of food. In two colonies of bilbies found near salt lakes, the bulbs of the sedge, *Cyperus bulbosus* (Cyperaceae), made up almost two-thirds of the faecal material. Elsewhere, the annual grass *Panicum australiense* (Graminae) contributed between one- and two-thirds of the identifiable fragments. Seeds of the trigger plant, *Stylidium desertorum* (Stylidiaceae), were also prominent in the faeces. Although bilbies, like bandicoots, are insectivore/omnivores, they appear to be much more dependent on plant material for food than the heathland bandicoots discussed previously.

Tarsipedidae

The single species in this family, *Tarsipes rostratus* (Honey possum) appears to feed exclusively on nectar and pollen (nectarivore). Duncan (1979) found no evidence of insects in the contents of alimentary tracts and faecal samples. This animal has a number of structural attributes which facilitate feeding on these resources (Rourke & Wiens, 1977). The snout is long and pointed and the mouth is tubular. The tongue is also long and extendible, and has brush-like papillae on its dorsal surface towards the pointed tip. A vascular sinusoidal mechanism facilitates extension of the tongue from the relaxed length (2 cm) to the extended length (4 cm), allowing penetration deep into flowers (Vose, 1973). There is also a diverticulum at the junction of the oesophagus and the stomach which may serve as an incubation chamber in which the pollen grains are germinated prior to digestion (Turner, 1984a). Honey possums probe the inflorescences of *Banksia* and *Eucalyptus* spp. and preen pollen from the fur between bouts of probing (Hopper & Burbidge, 1982).

T. rostratus is scansorial and crepuscular (Hopper & Burbidge, 1982), weighing between 7 and 20 g when adult. It is confined to heathland in south-western Australia, where a variety of proteaceous and myrtaceous species provide an abundant year-round supply of nectar and pollen (Wooller *et al.*, 1981).

Burramyidae

There are seven extant burramyids: six pygmy possums, which bear closest resemblance to certain didelphids (*Marmosa* spp.) and nocturnal lemurs (e.g. *Microcebus*; Primates), and the pygmy glider, *Acrobates pygmaeus*. They fall within the weight range 7–45 g (see Fig. 2.2 for body length distribution). The pygmy possums are mouse-like with prehensile tails and grasping feet. The pygmy glider resembles a dormouse, but has gliding membranes between the fore and hind limbs and a feather-like tail which has hairs growing from each side. Burramyids are nocturnal and mostly arboreal or scansorial, inhabiting forests and heaths. One species, *Burramys parvus* (see Plate 2.3) lives in rocky alpine screes and associated heaths (Gullan & Norris, 1981).

Cercartetus nanus is primarily a nectarivore, but also takes some fruit and insects. Turner (1984a) found that *C. nanus* harvests nectar and pollen from the large inflorescences of *Banksia* spp. (Proteaceae) and the flowers of *Leptospermum laevigatum* (Myrtaceae). The fruits of *Leucopogon parviflorus* (Epacridaceae) are eaten in summer. Remnants of beetles were found in the faeces in all seasons, but were most abundant in summer. Fragments of moths were most common in spring, especially during blooms of the Bogong moth (*Agrostis infusa*; Noctuidae). Insect larvae

Fig. 2.2. Distribution of body lengths (snout-vent; mm) of some marsupial families. (*a*) Burramyidae (fine stipple), exudate-feeding Petauridae (coarse stipple), folivorous Petauridae (open), Phalangeridae (cross-hatched), *Phascolarctos cinereus* (shaded); (*b*) Macropodidae (stippled), Vombatidae (open). For data see Appendix 1.

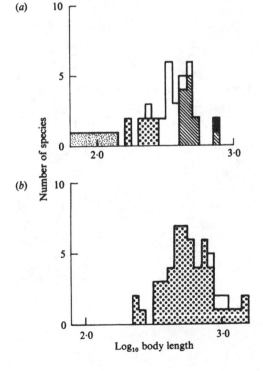

Plate 2.3. *Burramys parvus*, a frugivorous/granivorous inhabitant of rocky screes. (Tony Siddons.)

formed part of the diet in winter and spring. It seems likely that most of these insects, including the larvae, were harvested from the *Banksia* inflorescences.

It is widely assumed that *Acrobates pygmaeus* (see Plate 2.4) feeds upon insects, but an examination of the faeces of this species by Turner (1984b) suggests that it too is a nectarivore. Digested pollen (probably from

Plate 2.4. *Acrobates pygmaeus*, the smallest marsupial glider. (Penny Gullan.)

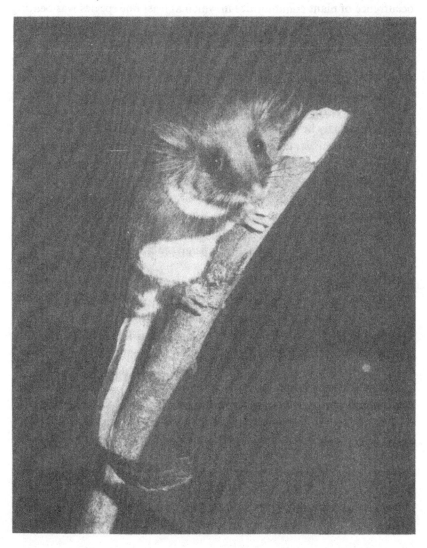

Eucalyptus spp.) was present in the faeces in most months. *A. pygmaeus* has been observed sucking sap from the sap sites on the trunks of *Eucalyptus* spp. (Myrtaceae) cut by *Petaurus australis* (Henry, 1984a), and may feed on other plant exudates such as manna and honeydew. Both *C. nanus* and *A. pygmaeus* groom pollen from their fur after bouts of feeding at inflorescences (Turner, 1984a, b).

Burramys parvus is found at only a few alpine localities in south-eastern Australia. Gullan & Norris (1981) found that captures of *B. parvus* during the snow-free period at Mt Hotham, Victoria, were correlated with the occurrence of plant communities in which at least one species was bearing fruit. In early summer captures were highest in the vicinity of a community containing fruiting *L. suaveolens* (Epacridaceae), and in late summer they were highest in the vicinity of a distinct community containing fruiting *Podocarpus lawrencei* (Podocarpaceae) and *Tasmannia xerophylla* (Winteraceae). The fruits of these shrubs are eaten by the possums in captivity (V. Turner, personal communication). The sweet-tasting fruit of *L. suaveolens* may be particularly significant for this possum as they persist on the plant over winter at a time when there is significant snow cover. *B. parvus* may also eat seeds and insects. M. Fleming (personal communication) observed captive animals extracting seeds of *Eucalyptus pauciflora* from mature seed capsules with the aid of their lower incisors. *B. parvus* will also cache seeds when maintained in captivity. This behaviour may be important in providing the animal with a food supply in winter. *B. parvus* appears best classified as a frugivore/granivore.

Petauridae

Twenty-three species of small- to medium-sized arboreal marsupials (0.12–1.7 kg; see Fig. 2.2 for body length distribution) occur in this family. They include Leadbeaters possum (*Gymnobelideus leadbeateri*), 5 gliding phalangers, which resemble 'flying' squirrels (Rodentia), having a gliding membrane between the fore- and hind-feet, 13 ringtail possums, similar in appearance to marmosets and lorises (Primates), and 4 striped possums. The striped possums have been likened to the Aye Aye (*Daubentonia madagascariensis*) of Madagascar with their chisel-like incisors and elongate fourth digit on the hand, which they probably use to extract insects from crevices in tree trunks. Most petaurids are nocturnal and arboreal forest dwellers. Local species diversity is highest in New Guinea and in the tropical rainforests of north-eastern Queensland.

The smaller petaurids such as *G. leadbeateri* and *Petaurus breviceps* feed primarily on plant exudates and insects. Identification of the exudates used by possums and gliders is fraught with difficulty since they are almost totally digested, and there is little or no trace in the faeces. A. P. Smith (1980, 1982a) and Smith & Russell (1982) used feeding observations and indicators of exudates in the faeces (e.g. bark fragments, which indicate sap feeding) to assess the importance of the various exudates, but these observations do not provide precise estimates of the biomass of exudates consumed. A. P. Smith (1980, 1982a) concluded that *G. leadbeateri* feeds upon a number of exudates including sap (manna) flowing from wound sites on *Eucalyptus regnans*, sugary solutions (honeydews) secreted by sap-sucking psyllids (Homoptera) and gums that flow from wound sites on the trunks and branches of *Acacia dealbata* and *A. obliquinerva* (Mimosaceae). Smith found that the principal arthropod prey of *G. leadbeateri* are tree crickets (Rhaphidophoridae), beetles, moths and spiders. Arthropods consistently represented in the faeces were tree crickets and spiders and these were available year-round under the decorticating bark of *E. regnans*, the dominant tree species in the habitat of this species. The occurrence of moths and beetle fragments peaked in the faeces during spring and summer.

Gum from *A. mearnsii* is important in the diet of *P. breviceps*, particularly in autumn and winter when it makes up the bulk of the diet. These gliders also gnaw through the bark of certain eucalypts (e.g. *E. bridgesiana*) to obtain sap, and also feed on nectar from eucalypt blossom (Smith, 1982b). While exudates are important food resources in winter, the gliders feed predominantly upon insects during spring and summer. In Smith's study (1982b) the most important insects in the diet were moths and scarabaeid beetles that had larval stages in pastures adjacent to the study area, and adults that were dependent upon eucalypts for shelter and food.

The largest of the plant-exudate/insect-feeding petaurids, *P. australis*, has an adult body weight of 0.43–0.53 kg, which is only slightly less than that of the herbivorous *Pseudocheirus* spp. Smith & Russell (1982) concluded that the bulk of the diet of this glider consists of exudates including eucalypt sap, eucalypt nectar and to a lesser extent honeydew. *Petaurus australis* also consumes a variety of arboreal arthropods including tree crickets, beetles, beetle and moth larvae, and spiders. Smith & Russell argued that *P. australis* probably relies more on plant exudates as sources of energy than the smaller *P. breviceps* because large size does not increase insect-harvesting potential. Insectivorous arboreal primates consume roughly the same quantity of insects regardless of body size (Charles-Dominique, 1974).

Smith & Russell (1982) found that in eucalypt forest in northern Queensland *P. australis* spent much of the time feeding at sap sites (see Plate 2.5). The gliders cut V-shaped notches in the bark of certain trees of the species *Eucalyptus resinifera* and attended these in all months. Elsewhere they cut similar notches in *E. viminalis*, *E. bridgesiana* and *E. racemosa* (Wakefield, 1970; C. Mackowski in Smith & Russell, 1982). At a study site in Boola Boola State Forest, Victoria, Henry (1984a) found that a colony of these gliders used purpose-made sap sites for only five months during a three-year period. Here most of the feeding time was spent foraging among loose bark high up in the crowns of trees, presumably in search of insects and manna. Nectar and pollen from the blossom of *Eucalyptus* spp. were also sought when available.

The exudate-feeding/insectivorous petaurids all nest in hollows in trees. They share nests and this has been demonstrated to reduce heat loss and so conserve energy (Fleming, 1980).

The diet of *Dactylopsila trivirgata* remains an enigma. Much has been made of the long procumbent lower incisors, the forward-projecting upper incisors and the elongate tongue and fourth fingers of this species, which are thought to be used to remove insects from cavities in the trunks of trees

Plate 2.5. Sap sites cut into the bark of *Eucalyptus* spp. by *Petaurus australis*. (Stephen Henry.)

(Rand, 1937). Smith (1982c) argued that it would be unusual for an animal of this size (0.30–0.35 kg) to feed predominantly upon solitary insects because of the high cost of harvesting, and pointed out that similarly-sized arboreal mammals include a significant amount of plant exudates in the diet. Two specimens collected by Mjoberg (Lonneberg & Mjoberg, 1916) contained lepidopteran and dipteran larvae in the alimentary tracts. These authors and Lumholtz (in Troughton, 1973) commented that this possum seeks out the communal nests of the small stingless bee *Trigona* sp. (Apidae) for both larvae and honey. Smith (1982c) examined the contents of the alimentary tract of two specimens and found a large quantity of ants and smaller quantities of termites and wood-boring beetle larvae, crickets, moth larvae and spiders. He noted the presence of termite nymphs and ant pupae and suggested that this indicates that *D. trivirgata* obtains some of its prey (termites and ants) by breaking into nests. These observations suggest that *D. trivirgata* feeds on both solitary and social insects.

In contrast to the foregoing, *Pseudocheirus* spp. are browsing herbivores. Thomson & Owen (1964) found that the stomach contents of *Pseudocheirus peregrinus* (see Plate 2.6) contained leaf fragments from species of *Eucalyptus*, *Leptospermum* and *Kunzea* (Myrtaceae), and from *Rubus fruticosus* and *Acacia dealbata*. Shoots and flowers had also been eaten. On any one night the diet may be less diverse; Thomson & Owen (1964) found remains of one plant species in 56% of stomachs, of two species in 40% of stomachs and of three species in only 4% of stomachs. Pahl (1984) has studied the diet and food selection of *P. peregrinus* in a series of plantations each containing different *Eucalyptus* spp. When offered a choice between leaves of different ages and species, *P. peregrinus* showed a preference for young foliage of *E. ovata*, *E. radiata* and *E. maculata* over old foliage of these species and foliage from understorey species such as *Rubus fruticosus* and *Leptospermum juniperinum*. *E. cladyocalyx* and *Cassinia longifolia* (Compositae) were avoided. The possums were in greatest abundance in plantations of the preferred eucalypts and these species contributed most to their diets. *P. peregrinus* nests in hollow tree trunks or constructs dreys from foliage in the branches of trees and shrubs.

The largest marsupial glider, *Petauroides volans*, also feeds upon eucalypt foliage (especially leaf tips), buds and flowers (Marples, 1973; W. J. Foley in Hume, 1982; Henry, 1984a). This species nests in hollow tree trunks.

Plate 2.6. *Pseudocheirus peregrinus*, an arboreal browsing possum. (Lester Pahl.)

Phalangeridae

There are eleven species of these arboreal phalangers, which bear some resemblance to lorises and bushbabies (Primates) and sloths (Edentata). They fall within the weight range 1–4.5 kg (see Fig. 2.2 for body length distribution) and are nocturnal and arboreal except for *Wyulda squamicaudata*, which is found in rocky escarpments and is scansorial. Phalangerids attain greatest species diversity in the tropical forest of New Guinea. *Trichosurus vulpecula* thrives in suburban environments in most Australian cities and has become both a pest and a commercially exploited species in New Zealand, where it was introduced early in the last century (Pracy, 1974).

Phalangerids are browsing herbivores. The open forest and woodland species *T. vulpecula*, and the closed forest species *T. caninus* both feed on plant foliage, but show differing dependence on eucalypt foliage and understorey species. Owen & Thomson (1965) found that understorey species contributed 20% of the stomach contents of *T. vulpecula* and 88% of the stomach contents of *T. caninus*. *T. vulpecula* had fed mainly on eucalypt foliage (see also Freeland & Winter, 1975) while *T. caninus* had fed on a variety of plants from the field layer, including herbaceous species, fungi and occasionally lichens. In northern Australia, *T. arnhemensis* may be frugivorous at certain times of the year (J. A. Kerle, personal communication). *Trichosurus* spp. nest in hollow tree trunks.

Phascolarctidae

The koala, *Phascolarctos cinereus*, is one of the largest arboreal browsers (6.5–12.5 kg). It is also unusual among browsing herbivores in the high specificity of its diet. It feeds almost exclusively on the foliage of *Eucalyptus* spp., and while it may at times feed off a variety of *Eucalyptus* spp., it shows a regional preference for one or two species. In Victoria, koalas prefer the leaves of *E. viminalis* and to a lesser extent *E. ovata*, while the preferred species in Queensland and northern New South Wales are *E. tereticornis* and *E. punctata* (Eberhard, 1972, 1978; Hindell, Handasyde & Lee, 1985). They occasionally feed on species in genera other than *Eucalyptus*, including the introduced *Pinus radiata* (Pinaceae) (Lithgow, 1982). The basis for this specificity is not yet understood.

Koalas rest in the forks of trees, and do not build nests. They are primarily nocturnal.

Macropodidae

There are about 56 extant species of macropods, and they are often placed in two subfamilies, the Potoroinae, embracing the small rat kangaroos (0.5–3.5 kg), and the Macropodinae, containing the kangaroos and wallabies (1.3–66 kg; see Fig. 2.2 for body length distribution). The geographic range of the family includes Australia, New Guinea and associated islands. The potoroines are semifossorial and terrestrial, and are found in a variety of plant communities ranging from rainforest to desert steppe. With the exception of the scansorial tree kangaroos (*Dendrolagus*), the macropodines are terrestrial. They occur in most plant communities within the range of the family.

The information we have on the diets of the Potoroinae suggest that they more closely resemble bandicoots than the macropodines. Fungal tissue, principally the fruiting bodies of hypogeous sporocarpic fungi, has been found in the diets of *Potorous tridactylus* (Guiler, 1971) and *Bettongia penicillata* (Kinnear, Cockson, Christensen & Main, 1979). *B. penicillata* is also known to feed on gum from a *Hakea* sp. Both species also feed on insects, and *B. lesueur* may even scavenge fish from the shore line (Ride & Tyndale-Biscoe, 1962). We view the Potoroinae as fungivores/omnivores. An exception may be *Hypsiprymnodon moschatus*, which eats both fruit and animal material, and can probably be classified as a frugivore/omnivore (Johnson & Strahan, 1982).

The Potoroinae are generally secretive and nocturnal, although Johnson & Strahan (1982) concluded that the rainforest species *H. moschatus* is diurnal with peaks of activity at dawn and dusk. Most appear to construct nests in dense vegetation, although *B. lesueur* nests in burrows (Ride & Tyndale-Biscoe, 1962). The nests are shared by several individuals (Stodart, 1966).

The radiation of the Macropodinae in many respects parallels the radiation of artiodactyls in Africa. They are all herbivores, filling a wide range of browsing and grazing niches. They also show convergence with artiodactyls in the solutions adopted for digestion of a diet rich in plant fibre and often poor in protein (Hume, 1982).

The large macropodines *Macropus giganteus*, *M. robustus* and *M. rufus* feed primarily on grasses which frequently contribute between 60 and 90% of the identifiable fragments of food in the stomach or faeces (Kirkpatrick, 1965; Griffiths & Barker, 1966; Chippendale, 1968; Storr, 1968; Bailey, Martensz & Barker, 1971; Bell, 1973; Griffiths, Barker & McLean, 1974; Dawson & Ellis, 1979). Dawson & Ellis (1979) found that grasses made up 61% of the diet of *M. robustus* even though grasses constituted only 1% of the ground cover at their study site.

Although *Themeda australis* (kangaroo grass) is the principal food item of *M. parma* (Maynes, 1974) and *M. parryi* (Bell, 1973), smaller macropodines appear to be less dependent upon grasses. For example, Copley & Robinson (1983) found that grass, forbs, dicotyledonous vegetation (browse) and plants with stellate trichomes all contribute significantly to the diet of *Petrogale xanthopus*, but none is consistently dominant. Grasses and forbs usually make the greatest contribution in winter and spring, when they are most abundant, browse is important in dry months, and plants with stellate trichomes are significant only in summer and early autumn, when fragments of the fruit of *Solanum petrophilum* (Solanaceae) are evident in most samples. A comparison between the composition of faeces and the availability of food plants suggested a weak preference for grass over other plant types, and that these wallabies are opportunistic.

In contrast with the foregoing, the diets of *Wallabia bicolor* (see Plate 2.7) and *Setonix brachyurus* show a predominance of browse. Edwards & Ealey (1975) found that *W. bicolor* eats a wide range of native and exotic vegetation, preferring the coarse browse provided by shrubs and bushes. On Rottnest Island, Western Australia, *S. brachyurus* feeds predominantly upon succulents in areas where surface water is never available. In a study by Storr (1964), succulents contributed 79% of the diet in spring and 97% of the diet during the hot dry summers. Even where water was available, succulents still made up 52–77% of the diet. Grass never comprised more than 31% of the diet, and herbs and grasses only assumed importance in winter and early spring, when annual species were responding to winter rainfall.

The majority of the macropodines appear to have no requirements for nest sites other than shade, but some such as *M. robustus* and the rock wallabies (*Petrogale* spp., *Peradorcas concinna*) use rock piles and caves. Most of the small species are nocturnal, whereas the larger macropods are generally nocturnal or crepuscular.

Vombatidae

There are three extant species of wombats, all broad-headed and stout-bodied terrestrial mammals with short, robust limbs. They fall within the weight range 20–35 kg.

Wombats are found in savannah woodlands, desert steppes and open eucalypt forest in eastern and southern Australia, including Tasmania. They all use burrows as nest sites, and for escaping from high daytime temperatures. *Vombatus ursinus* individuals tend to live alone in burrows and use a number of burrows scattered through their home range. Burrows of *Lasiorhinus latifrons* may occur in close proximity to one another, forming large warrens. The location of burrow systems may be influenced by breaks in subsurface calcrete, but there is no evidence that these sites limit the number of burrows (Wells, 1978).

There have been no systematic analyses of the diet of wombats, although they are known to feed on grasses and possibly roots. Wombats are nocturnal.

Plate 2.7. *Wallabia bicolor*, a browsing macropod. (Bruce Fuhrer.)

Feeding categories and the quality and dispersion of food

The feeding categories discussed below are adopted from Eisenberg (1981) and summarised in Table 2.3.

Carnivore

Vertebrate prey offers predators the advantages of easily assimilated food with a composition close to the predators' requirements. Predators able to capture mammals and non-migratory birds are ensured a year-round food resource, although this may be locally depleted where the prey species breeds seasonally and infrequently. An additional advantage of vertebrate prey is the relatively large size of most prey items (cf. insects), but their mobility may increase the cost of harvesting. One means of reducing this cost, found in the Tasmanian Devil (*Sarcophilus harrisii*), is to feed upon vertebrate carrion.

One of the curiosities of Australian mammalian fauna is the paucity of truly carnivorous species (Keast, 1971). In the Recent fauna only the eutherian dingo (*Canis familiaris*) and the marsupials *Thylacinus cynocephalus* and *S. harrisii* fall clearly into this category, and they only occupy the weight range 6–25 kg. Yet the marsupial radiation was clearly capable of producing a diverse array of predatory forms as witnessed by the extinct Borhyaenidae of South America. Explanations of this phenomenon have usually centred upon the inadequacies of the biomass of herbivorous mammals in Australia in providing an adequate environment for the evolution of marsupial carnivores (Keast, 1971, 1972; Janzen, 1976). It is interesting to note that the last stronghold of the large dasyurids is Tasmania, which is one of the most climatically predictable areas of Australia. The dependence of carnivorous marsupials on reliable food resources is reflected in the increase in numbers of *S. harrisii* in Tasmania since settlement, apparently in response to the increased availability of carrion (Guiler, 1970a).

Insectivore/omnivore

Mammals in this category feed primarily upon arthropods, especially insects. While arthropods offer similar advantages to vertebrate prey in terms of quality, they often show marked seasonal variation in abundance. This variation is particularly evident among flying and arboreal insects in most temperate communities. The biomass of flying insects may be an order of magnitude higher in summer than in winter (A. P. Smith, 1980). During winter there is a paucity of large flying

Lepidoptera and Coleoptera, but sometimes an abundance of small diptera (A. P. Smith, personal communication). Leaf- and shrub-dwelling insects also tend to peak in abundance in spring and summer. The value of these arthropods as a source of food depends upon the rate at which they can be harvested and their rate of renewal. Most have seasonal life cycles and so are only available for a short period. Once harvested, they may not become available again for many months. Insectivorous marsupials generally appear to be opportunistic, resolving these problems by turning to other life history stages (e.g. the larval stage) or to arthropods which show little seasonal change in abundance, when arboreal or flying insects are uncommon. A. P. Smith (1980) found that arthropods sheltering beneath the bark of trees in woodland peaked in abundance in winter, and that the number of leaf-dwelling spiders changed little with season. Soil insects may also provide a useful food resource in autumn and winter. Many of the arboreal insects which are abundant on the foliage in spring and summer have larval stages which occur in the soil during autumn and winter, and these are capable of providing a substantial food resource for animals like bandicoots which are able to harvest the larvae.

Small insectivores (< 0.1 kg) appear to be able to subsist almost entirely on arthropods, but may take fruit and flowers when these are locally abundant. The diets of large insectivores, however, usually include fruit, vertebrate prey or fungal sporocarps. These marsupials may be forced towards omnivory by the high cost of harvesting generally small arthropods.

At this time our knowledge of the diets of marsupials in this category is too limited to define the size limits of marsupial insectivory and insectivory/omnivory. Together they cover the weight range 0.003–4.7 kg.

Myrmecophage

In view of the abundance of ants and termites in terrestrial communities in Australia, it is surprising that marsupials have been reticent in utilising these resources. Only one marsupial, *Myrmecobius fasciatus* is clearly myrmecophagous, and shows the general array of structural characters associated with this feeding mode in mammals (see above, Myrmecobiidae). Large mammalian insectivores all feed on social insects (Kay & Hylander, 1978), and in doing so must reduce the cost of harvesting insect prey. A nest of social insects may be regarded as a single large organism. The failure of marsupials to utilise the rich and abundant Australian ant fauna is probably related to the ants' unpalatability.

Exudate feeder/insectivore

Most of the foregoing insectivores are scansorial, terrestrial or semifossorial and harvest a proportion of their food from the soil or ground surface. We have suggested that this provides a more stable year-round resource than arboreal and flying insects. Arboreal marsupials resolve the problem of seasonal variation by supplementing their diet with plant or insect exudates. These exudates include sap, manna, honeydew and gum.

The saps of plants are rich in soluble sugars (Stewart *et al.*, 1973), but are generally inaccessible to vertebrates because in most plants sap does not usually flow when the phloem ducts are mechanically severed (Canny, 1973). However, a variety of homopterous insects (cicadas, scale and lerp insects, aphids and leaf hoppers) have succeeded in exploiting this resource. These insects pierce rather than sever the ducts, and frequently imbibe large quantities of sap, extracting the nitrogenous substances from it and secreting excess unwanted sugars as honeydew. Honeydew-producing insects may rely on ants, birds and possibly mammals to remove excess secretions, otherwise these can accumulate, attract moulds and kill the insect (Matthews, 1976). Sap may flow from the wound sites deserted by sap-sucking and leaf-eating insects and leave a white encrustation of sugars called manna. Manna production in some eucalypts is substantial and can exceed nectar production at certain times of the year (Paton, 1979).

Honeydew and manna production are seasonal and presumably fluctuate with the population of sap-sucking insects, which reaches a peak in spring and autumn in temperate eucalypt forests. Manna and honeydew may be seasonally important in the diets of the petaurids *Petaurus breviceps* and *Gymnobelideus leadbeateri* (A. P. Smith, 1980), and probably in those of *P. australis* and the burramyid *Acrobates pygmaeus*. *P. breviceps* gleans manna from eucalypt foliage and licks honeydew from the smooth branches of eucalypts where it is produced by naked lerp insects.

A small number of trees produce sap after mechanical injury of their phloem ducts (Canny, 1973), and prominent among these are a number of *Eucalyptus* spp. We have seen that *P. australis* makes V-shaped incisions in the trunks of certain eucalypts and feeds at these sites. The incisions also attract *P. breviceps*, *A. pygmaeus*, *Antechinus stuartii* and *Phascogale tapoatafa* (Wakefield, 1970; Russell, 1980; Henry, 1984a). *Petaurus breviceps* also chews holes in the trunks of *Eucalyptus bridgesiana* and is able to obtain sap from these holes (A. P. Smith, 1980). The flow from sap sites is generally very slow and animals using this resource commonly spend long periods of five or more hours at each site (Henry, 1984a). It is still not clear why only certain trees of certain species produce sap after mechanical damage to their phloem ducts, or why this resource is available or used only at certain times.

Gums, like manna, may be produced in response to insect damage to certain tree species. *Acacia* spp. produce gum during the hottest and driest periods of the year. Most gums are non-toxic carbohydrate polymers which require microbial degradation if they are to be digested (Adrian, 1976). They also absorb large quantities of water and so can greatly increase the animal's water requirements. *Acacia* gums differ in their solubility and this influences their availability. The gum of *Acacia mearnsii* is produced only in spring and summer, but is insoluble in water and may persist on the trees for up to twelve months after production. Such gum provides an important winter food resource for *P. breviceps* (Smith, 1982b). The gums of *A. obliquinerva* and *A. dealbata* are produced in greatest abundance in summer and autumn, but are soluble in water, and so are not available to *G. leadbeateri* as a winter food resource. Kinnear *et al.* (1979) suggested that the gum from a *Hakea* sp. eaten by *Bettongia penicillata* provides an energy substrate for anaerobes in an expanded foregut. They considered that these anaerobes are essential for the digestion of the principal dietary item, fungal sporocarps. However, microbial digestion in the foregut may also be necessary for utilisation of this gum.

Most exudates are rich in carbohydrates but poor in protein. Consequently all marsupial exudate feeders include insects or fungal sporocarps in their diet. Even when arboreal insects were scarce, Smith (1982b) found that *P. breviceps* spent a considerable part of each night searching for arthropods.

Marsupial exudate feeder/insectivores fall into the weight range 0.12–0.53 kg, which is similar to that of prosimians within this dietary category (Hladik, 1979). These marsupials may be too large to subsist off insects alone, especially when these are highly seasonal, and turn to gum and other exudates as readily harvestable sources of energy. The size of exudate feeder/insectivores may be determined by the cost of harvesting these small and often scattered food resources.

Fungivore/omnivore

The nutritive value of fungal tissue is poorly understood. Some tissues have a calorific value which compares favourably with that of fruit but is less than the calorific values of seed and animal tissues. Fungal tissue is a potentially valuable source of protein and carbohydrates, especially for animals capable of digesting structural carbohydrates. The nutritional value of fungal tissue most closely resembles that of foliage among the tissues of higher plants, but fungi lack tannins, which in higher plants complex with proteins to reduce their digestibility (Fogel & Trappe, 1978; Martin, 1979).

Many fungi have above-ground fruiting bodies (epigeous sporocarps) and spores that are dispersed by the wind, and many of these fungi are toxic. Others have subterranean fruiting bodies (hypogeous sporocarps) that are attractive to small mammals and dependent upon them for dispersal (Maser, Trappe & Nussbaum, 1978). Hypogeous sporocarps appear to occur almost universally in the diets of small eutherians (Maser *et al.*, 1978). They are a potentially rich source of lipids and nitrogen but imbalanced in certain amino acids (Kinnear *et al.*, 1979). Their availability may be restricted to the wetter months in some seasonally dry Australian communities (Cockburn, 1981a). The seasonality of this resource and its imbalanced nutritional value are probably the reasons why marsupials feeding upon fungus also feed upon insects and gum.

The extent of the use of fungal sporocarps by marsupials has not been fully ascertained. The two potoroines known to feed extensively on this resource occur in the weight range 1020–1300 kg.

Frugivore/omnivore

Fruits and fleshy pericarps of many plants are attractive to birds and mammals and are presumed to have evolved as a means of disseminating seeds (McKey, 1975). They are generally rich in non-structural carbohydrates or lipids but poor in protein content. Toxic compounds are generally absent from ripe fruit, although they may protect immature fruits. Some fruits have laxative effects, presumably to ensure rapid deposition of seeds. The phenology of fruiting is directly linked to the phenology of flowering, and neither should be discussed in isolation. This topic is reviewed in some detail in Chapter 7. Most tropical Didelphidae feed on fruit when it is available, and members of the subfamily Caluromyinae are probably frugivorous throughout the year. Several species of plants in neotropical forests appear to be dispersed predominantly by marsupials, and have undergone selection to promote frugivory by didelphids (see Chapter 7). However, because fruit is only seasonally available and is usually poor in protein, these frugivores usually supplement their diets with exudates and insects.

Fleshy fruits are generally scarce in temperate eucalypt forests and woodlands in Australia, but are found more often in tropical forests

(Clifford & Drake, 1980). They are also available in some heathland and scrubland communities where genera such as *Styphelia, Astroloma, Leucopogon* (Epacridaceae) and *Coprosma* (Rubiaceae) produce edible fruits, some of which are eaten by marsupials (e.g. *Cercartetus nanus*; V. Turner, personal communication). However, fruit production by these genera is usually seasonal and where several species occur together, they rarely provide a year-round resource. No Australian marsupial appears to be strictly frugivorous although some, such as *Pseudocheirus peregrinus, Burramys parvus, C. caudatus, C. nanus, Trichosurus arnhemensis* and *Hypsiprymnodon moschatus*, and even the largely insectivorous *Antechinus* spp., feed on fruit when it is available. The opportunities for frugivory are substantial in the tropical rainforests of Australia and New Guinea, and both *H. moschatus* (Johnson & Strahan, 1982) and *T. arnhemensis* (J. A. Kerle, personal communication) have recently been shown to be partly frugivorous in these habitats. We have little or no information on the diets of other marsupials inhabiting those rainforests.

Frugivore/granivore

Seeds generally contain stored nutrients that make them highly nutritious and suitable as food. Seed loss to predators is obviously disadvantageous to plants, and they have accordingly developed a number of different mechanisms for reducing predation. These include the production of thick and heavy or toxic seed coatings (Janzen, 1971), of large quantities of small, edible seeds which are produced in short, intense bursts that saturate predator requirements (Janzen, 1971) and of eliasomes (ant-attracting substances) that favour the dispersal and burial of seeds by ants, so removing them from the site of potential predation (Berg, 1975). The latter strategy is particularly common in Australia, where more than 1500 species in 87 genera have seeds with eliasomes that are dispersed by ants (Berg, 1975).

There appear to be no specialist granivores among marsupials. Perhaps the most seed-dependent species is *B. parvus*, which may cache seeds as a means of providing food beneath the snow during winter. We have seen that the principal dietary item of this species is probably fruit.

Morton (1979) suggested that the general low diversity of granivorous mammals in Australian deserts, as compared with that found in North American deserts, was due to competition for seeds by birds and ants. Some endemic rodents feed upon seeds, particularly when these are abundant in late spring and summer, but none appears to rely exclusively upon this resource (Watts & Braithwaite, 1978; Cockburn, 1981b). There is no evidence that these rodents have excluded marsupials from this niche.

Nectarivore

Plant components and products vary considerably in their nutritional quality. Nectar, for example, is rich in simple sugars but contains little or no protein, whereas pollen is rich in amino acids, although these may be difficult to extract from the pollen grain (Turner, 1984a). The inflorescences of a number of Australian plants, such as *Banksia* spp., produce copious quantities of nectar and pollen which serve to attract bird and mammal visitors. The coevolutionary relationships between these plants and marsupial pollinators is discussed in Chapter 7.

The value of nectar and pollen to marsupials will depend upon the seasonality of flowering, the amount of nectar and pollen available per inflorescence, the abundance of inflorescences and the rate at which inflorescences are visited. Marsupials will be able to specialise in exploitation of nectar and pollen only where individual plant species have prolonged flowering times or where the marsupials occur in a community where the plant species have staggered flowering times. The most specialised of all of the marsupial nectar and pollen feeders, *Tarsipes rostratus*, is confined to floristically-rich heathland communities in south-western Australia, where these resources are available in all months (Wooller *et al.*, 1981). Although *C. nanus* feeds off nectar and pollen when *Banksia* spp. and *Leptospermum laevigatum* are flowering in plant communities on Wilson's Promontory, Victoria, it turns to other resources, particularly fruit, during the non-flowering season (Turner, 1984a).

Nectarivorous marsupials are uniformly small in size (7–30 g), and so can be expected to have high metabolic rates and energy requirements.

Herbivore

Although leaves are rich in carbohydrates, much is structural, consisting of cellulose, hemicellulose and lignins, which are not digested by mammals lacking cellulases (Van Soest, 1982). These structural carbohydrates make up the cell walls. The other distinct fraction in leaves is the cell content. This contains lipids, sugars, starch, soluble proteins, vitamins and amino acids, and may be almost totally digested by mammals. The nutritional value of leaves to mammals depends upon the proportion of cell wall to cell content and the animal's ability to digest the cell wall and to crack the cell wall to gain access to the cell content. Mammalian herbivores use mechanical digestion by the teeth and a culture of bacteria which possess cellulases to achieve these objectives (see Hume, 1982).

Most Australasian tree communities are evergreen and provide a continuously-available resource of mature foliage. However, the quality of leaves may change dramatically with age. A bite from a young leaf will include many thin-walled cells, while a similarly-sized bite from a mature leaf will include fewer, larger cells with thicker walls, little cytoplasm and large vacuoles filled with polyphenols. The secondary compounds which accumulate in these vacuoles may be toxic or chelate proteins and so render their digestion difficult (Freeland & Janzen, 1974). Young leaf may contain more or similar amounts of structural carbohydrates than old leaf of the same species because the same area of young leaf contains more cell walls than old leaf. But of greater importance is the fact that young leaf may contain more cytoplasm and therefore protein. Parra (1978) considered that the low digestibility and low protein content renders many mature plant leaves nutritionally suboptimal for mammals, which may only be able to gain adequate nutrition from the more readily-digestible younger leaves and buds. Clearly the phenology of the food plants available to herbivores will largely determine the quality of their diet.

Seasonal changes in the quality of leaves are even greater in grasses, where, during the non-growing season, proteins and carbohydrates may be withdrawn into the roots leaving a highly lignified and relatively indigestible remnant leaf with low protein and energy content (Sinclair, 1975). However, unlike the foliage of many trees and shrubs, grasses rarely store toxic or protein-chelating compounds. For example, alkaloids and cyanogenic glycosides are only known to occur in about a hundred of the eight thousand or so species of grasses (Culvenor, 1970).

With the exception of rodents, herbivory is confined to mammals which attain an adult body weight of 0.6 kg or more. It is generally assumed that this size constraint is a consequence of the low readily-accessible energy available from leaves and the need for mammalian herbivores to slow the passage of food in order to permit bacterial degradation of the structural carbohydrates. Further, it is widely accepted that small herbivores, with a limited gastrointestinal size and therefore limited food intake and storage, need to be highly selective in their feeding, selecting items which are rich in energy and protein (Jarman, 1974; Van Soest, 1982). It is this requirement which is thought to restrict small herbivores to browse, whereas large herbivores with correspondingly greater energetic require-ments tend to be grazers. Although the large diprotodont marsupials tend to be grazers, the expected size division between browsing and grazing diprotodonts is not clear. Some of the small *Macropus* spp. feed primarily upon grasses.

The missing marsupials

Eisenberg (1981) defined 53 macroniches available to mammals according to their different needs of food and substrate (Table 2.4). Only 15 of these macroniches are clearly filled by marsupials. We have already noted the lack of aquatic and volant marsupials, and the paucity of fossorial, semiaquatic, carnivorous and granivorous marsupials. In addition, strictly diurnal species are restricted among marsupials to a single species, *Myrmecobius fasciatus*.

Fig. 2.3. Distribution of average body lengths (snout-vent; mm) for (*a*) marsupial and (*b*) all mammalian genera. From Eisenberg (1981).

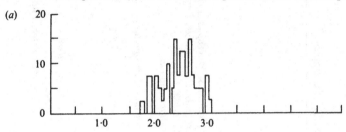

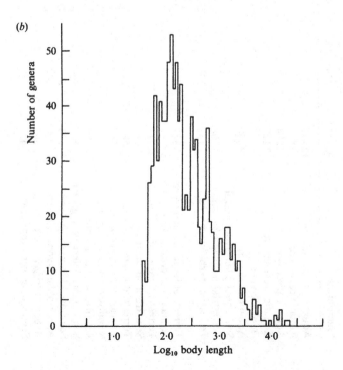

Table 2.4. *Mean body lengths for all mammalian genera and for all marsupial genera separated into fifty-five 'macroniches'. Genera classified according to modal tendencies*

Macroniche	Mean body length[a] (mm)		Marsupial examples
	Mammals	Marsupials	
Aquatic planktonivore	12491		
Aquatic carnivore	4875		
Aquatic piscivore	3204		
Aquatic crustacivore/molluscivore	3127		
Aquatic herbivore	3050		
Semiaquatic herbivore/grazer	1870		
Terrestrial herbivore/grazer	1508	550	*Lagorchestes, Macropus, Peradorcas, Petrogale, Onychogalea*
Terrestrial herbivore/browser	1132	535	*Dorcopsis, Dorcopsulus, Lagostrophus, Setonix, Thylogale, Wallabia, Wyulda*
Terrestrial frugivore/omnivore	817	236	*Hypsiprymnodon*
Semiaquatic crustacivore	737		
Terrestrial carnivore	718	881	*Thylacinus, Sarcophilus*
Semiarboreal carnivore	597		
Terrestrial myrmecophage	585	245	*Myrmecobius*
Scansorial myrmecophage	575		
Semifossorial myrmecophage	573		
Semifossorial carnivore	570		
Semifossorial herbivore/grazer	565	956	*Lasiorhinus, Vombatus*
Scansorial herbivore/browser	533	692	*Dendrolagus*
Arboreal herbivore/browser	508	469	*Petauroides, Phalanger, Phascolarctos, Pseudocheirus, Trichosurus*
Arboreal frugivore/herbivore	501		
Semiaquatic herbivore/browser	498		

Scansorial frugivore/omnivore	477		
Arboreal frugivore/omnivore	473	221	*Caluromys, Caluromysiops, Glironia*
Terrestrial frugivore/granivore	445	112	*Burramys*
Arboreal carnivore	427		
Semiaquatic piscivore	321		
Terrestrial fungivore/omnivore	309	319	*Aepyprymnus, Bettongia, Potorous*
Scansorial frugivore/herbivore	301		
Terrestrial frugivore/herbivore	289		
Semiaquatic insectivore	—		*Chironectes*
Semifossorial herbivore/browser	227		
Arboreal frugivore/granivore	265		
Semiaquatic frugivore/omnivore	252		
bArboreal gumivore	248	195	*Gymnobelideus, Petaurus*
Terrestrial insectivore/omnivore	223	196	*Antechinus, Caenolestes, Chaeropus, Dasycercus, Dasykaluta, Dasyuroides, Dasyurus, Echymipera, Isoodon, Lestodelphys, Lestoros, Lutreolina, Macrotis, Metachirus, Microperoryctes, Monodelphis, Murexia, Myoictis, Neophascogale, Ningaui, Parantechinus, Perameles, Peroryctes, Phascogale, Phascolosorex, Planigale, Pseudantechinus, Rhyncholestes, Rhynchomeles, Satanellus, Sminthopsis*
Scansorial insectivore/omnivore	221	285	*Didelphis, Dromiciops, Philander*
Fossorial herbivore/browser	209		
Fossorial frugivore/herbivore	202		
Arboreal insectivore/omnivore	198	164	*Dactylopsila, Distoechurus (?), Marmosa*
Semiaquatic frugivore/granivore	178		
Semiaquatic frugivore/herbivore	175		
Scansorial frugivore/granivore	165		

Table 2.4 (cont.)

Macroniche	Mean body length[a] (mm)		
	Mammals	Marsupials	Marsupial examples
Arboreal myrmecophage	156		
Semifossorial frugivore/ herbivore	144		
Arboreal nectarivore	118	75	*Acrobates, Cercartetus, Tarsipes*
Fossorial insectivore/omnivore	117	140	*Notoryctes*
Semifossorial frugivore/ granivore	104		
Volant carnivore	104		
Volant frugivore/omnivore	101		
Semifossorial insectivore/ omnivore	88		
Volant sanguivore	81		
Volant nectarivore	75		
Volant piscivore	74		
Volant foliage-gleaner/ insectivore	66		
Volant aerial insectivore	58		
Median length for mammalian/marsupial genera	619	250	

Macroniches for all mammals after Eisenberg (1981). The categories Terrestrial fungivore/omnivore and Semiaquatic insectivore are added in our analysis.

[a] Mean length of generic averages for each macroniche.
[b] We have used the term exudate feeder for gumivore in the text.

We have also pointed to the limited size range of marsupials in some of these macroniches (e.g. terrestrial carnivores) (see also Table 2.4 and Fig. 2.3). Some of these apparent limitations to the marsupial radiation obviously refer only to extant forms and it is important to remember that extinct marsupials include large carnivores (e.g. Borhyaenids), large herbivores (e.g. Diprotodont spp.), potential desert granivores (e.g. Argyrolagidae) and even marsupial 'leopards' (*Thylacoleo* spp.).

3

The marsupial life history

In this chapter we consider what it means to be a marsupial. We do this by contrasting the scope of the marsupial and eutherian radiations, and examine hypotheses which attempt to explain why the marsupial radiation appears to be conservative. It is not our intention to enter into the sterile debate over the general advantages of 'marsupialness' and 'eutherianness'. The coexistence of these groups in South America and Australia clearly testifies to the evolutionary viability of both.

However, we feel it is useful to point to a number of deficiencies in previous attempts to contrast the two taxa. Historically, it has often been assumed that marsupials are in some sense inferior to eutherians (Asdell, 1964; Lillegraven, 1975, 1979), perhaps as a consequence of the widespread but fallacious belief that marsupials represent an intermediate grade of mammalian organisation between monotremes and eutherians. Both the original work and attempts to destroy the assumption (e.g. Kirsch, 1977a, b; Parker, 1977; Low, 1978), suffer from a lack of quantification and statistical examination, and have given rise to summary statements which are probably incorrect, for example, 'the nonseasonal opportunism in most marsupials' (Low, 1978, p. 206), or 'even modest territorial behaviour in the marsupials is a comparative rarity' (Lillegraven, 1979, p. 270).

There is a further underlying and unrecognised assumption which pervades this literature. This may be paraphrased: all aspects of an organism are perfectly tuned or adapted to their environments. As we shall see, the case for interpretation of species characteristics in an adaptive framework is by no means always clear, and the application of similar assumptions to the characteristics which distinguish higher taxa is even less justified.

In order to circumvent these difficulties, it is necessary to come to terms with the concepts of fitness and adaptation, and to introduce a statistical procedure for the comparison of higher taxa. Although fitness of individual organisms is the cornerstone concept around which Darwin's theory of natural selection is based, attempts at definition of fitness have often led to circularity. Most evolutionary biologists agree that fitness will ultimately be expressed either in increased survival or in fecundity. Unless the phenotype of an organism confers increased survival or reproductive potential, and expression of that phenotype has a genetic basis, natural selection will not operate. For this reason, increased attention has been paid to variation in life history traits such as age at reproductive maturity, the number of offspring, the number of reproductive events per season and lifetime, reproductive effort and survival and longevity. Life history theory stems from Cole's (1954) classic examination of variation in these para-meters, and is now one of the most rapidly expanding areas of evolutionary biology (Stearns, 1980). Unfortunately, much of this theory is based on complex mathematics that is inaccessible to most biologists. As Horn (1978, p. 411) comments: 'Many of the important papers are unintelligible even to the authors of other important papers'. Despite this, we are convinced that a knowledge of life history theory is essential for under-standing both ecology and evolutionary biology, and variation in life history traits is the subject of the following four chapters.

Adaptation and constraint

Many biologists embrace the concept of adaptation as the con-sequence of evolution by natural selection. Organisms and traits of organisms are believed to fit precisely the environment in which they live, and the attention of ecologists and other biologists has become increasingly directed towards explanations invoking adaptation. This viewpoint has been reinforced by the mathematical procedure most commonly applied to life history theory, optimality theory (see Tuomi & Haukioja, 1979a, b), which seeks the mathematically-optimal solution to a biological problem. The utility of this approach has been widely criticised (Stearns, 1976; Oster & Wilson, 1978; Lewontin, 1979), particularly because of constraints which prevent organisms from attaining the optima predicted by mathe-matics (Stearns, 1977). For example, the ultimate optimal 'Darwinian demon' would start to produce offspring immediately after its own birth, and continue to produce very large numbers of young at frequent intervals while experiencing no mortality, impediments to dispersal or difficulty in finding food or mates (Law, 1980). Clearly, such an organism does not exist.

For these and more complex reasons, a number of authors have recently cautioned against the assumption of precise adaptation, and have stressed the difficulties of distinguishing between several plausible adaptationist explanations (Gould & Lewontin, 1979; Harper, 1982). Gould & Lewontin (1979) draw a telling analogy between the extreme application of adaptationist assumptions in biology and the absurd expostulations of Voltaire's fictional character Dr Pangloss, who believed we live in the best of all possible worlds and thus the function of any phenomenon defined both its purpose and the reason for its origin.

Harper (1982) has catalogued some of the various forces which influence the evolution of organisms and might prevent the attainment of precise adaptation. These factors may be usefully classified into two distinct classes (Stearns, 1982). These are:

(1) Constraints, or features of ontogenetic mechanisms and morphogenetic design which limit the power of selection to mould phenotypic traits. For example, the complex genetic and physical organisation of ancestral organisms may place limits on the scope for new changes. This may be particularly important when new populations are established because the founders of populations (e.g. a gravid female) may have a restricted gene pool which need not necessarily correspond to the average genotype of the source population (founder effects). Another important constraint on optimality arises where different traits of organisms vary conjointly, restricting the possibility of some traits being affected by separate evolutionary processes. The sources of such conjoint variation may be (i) genetic, through pleiotropism and linkage disequilibrium, (ii) allometric, as changes in the size or growth of a character of an organism are rarely confined to a character in isolation but are expressed simultaneously in many properties of the organism, and (iii) epigenetic, with development canalisation restricting the range of phenotypes which may be expressed in an organism (see e.g. Alberch, 1980). These factors led Gould & Lewontin (1979) to promote an alternative viewpoint which is prevalent among European biologists, and interprets the function and origin of traits in terms of the development of the entire organism, or *Bauplane*, rather than assuming precise adaptation of individual traits.

(2) Trade-offs, or compensation, which have the characteristic that if opposing selection forces were removed, the phenotype would be free to move beyond the point already attained (Stearns, 1977). For example, increased energy expenditure on reproduction may

restrict the energy available for growth or maintenance, an interaction we analyse in Chapter 6.

Even selection, like the evolutionary forces described above, need not result in an 'optimal' or precise fit between organism and environment. It is important to realise that many of these factors causing change and restrictions to change will contribute to most traits of organisms. Past events are also crucial in the study of evolution. If we are to understand the origin and performance of a trait it is inadequate to consider just its present function. It is also important to consider its origin and the evolutionary forces which constrain and narrow the range of activities of organisms. As Harper (1982, p. 17) states:

> This means that rather than concentrating on a search for the ways in which organisms are perfectly suited to their environments, we might more healthily concentrate on the nature of the limitations that constrain where they live. We might usefully ask not what is it about an organism that enables it to live where it does, but what are the limits and constraints that prevent it living elsewhere.

And Stearns (1982, p. 237) has contended:

> Organisms appear to be a mosaic of relatively recent adaptations which we can understand in terms of optimality theory, embedded in a framework of relatively old constraints which we would like to understand in terms of developmental mechanisms.

This chapter considers the topic of constraint and in the following chapters we discuss variation within those constraints, or the set of adaptations which characterise the extant marsupials.

Allometry

Before we proceed further it is necessary to digress briefly and introduce the allometric methodology we shall use throughout this chapter. Readers wishing to avoid this discussion should proceed to p. 50. Many morphological, physiological and even behavioural features vary among individuals and species as a consequence of differences in body size. Allometry is the study of size and its consequences (Gould, 1966). Most contemporary treatments stem from Huxley's (1932) suggestion that the power function

$$y = bx^k \qquad (3.1)$$

where x is body size, b and k are constants and y is a size-related variable, represents a general biological relationship relating growth of different parts of organisms. In provocative reviews of application of the power function, S. J. Smith (1980, 1981) argues that several unrecognised assumptions are contained in the following widely applied technique for interpreting allometric relationships:

(1) Transform variables to logs.

(2) Calculate a Pearson product-moment correlation coefficient to indicate the strength of the bivariate relationship.

(3) Provide a graph of plotted points to supplement the correlation as a visual demonstration of goodness of fit.

(4) Interpret the results by discussing the biological implications of the observed slope and/or explain the biology of individuals or species which fall furthest from the regression line.

S. J. Smith (1980, 1981) points out that the goodness of fit of the power function is only rarely compared with other functions, despite Thompson's (1961) demonstration that untransformed linear equations fit several data sets just as well as power functions. Analysis of thirty interspecific and thirty intraspecific data sets suggests that power functions perform consistently better only for interspecies data (S. J. Smith, 1980; Harvey, 1982). Further, this improved performance may be related to the way in which a log transformation reduces the problem of outliers, and can alter the distribution of data to meet statistical assumptions of normality and homoscedasticity. Unfortunately, these properties also reduce the capacity of human perception to detect outliers, especially given the curved nature of the confidence limits of a regression equation.

S. J. Smith (1981) also demonstrates that the value of the correlation coefficient is sensitive to the range of variables on the x- and y-axes, meaning that comparisons of the strength of the correlation coefficient are not always suitable for distinguishing which of two allometric relationships has the greatest biological significance.

A further problem with the use of linear regression as a statistical descriptor of the power function stems from the difficulty in distinguishing the dependent and independent axes in biological examples. As both axes will generally be measured with error, the choice of independent variable will influence the values of allometric constants (Harvey & Mace, 1982). Although alternative procedures which minimise this difficulty are available (see e.g. Kermack & Haldane, 1950; Harvey & Mace, 1982), associated statistics for the comparison of different power functions are poorly understood (Harvey & Mace, 1982), so we have retained the linear regression model for this study. Body mass (in kg) is used as the independent variable throughout, and where values of the dependent variables were not taken from specimens of known mass, the reference values listed in Appendix 1 were used. Eutherian and marsupial curves were compared first for differences in slope (k), and, if no difference was found, for differences in elevation (b), using the tests described by Zar (1974). To avoid interruption of the text, statistics of comparison are presented in Appendix 2. Where curves are statistically different, the difference is expressed using the allometric cancellation procedure described by Stahl (1962). By way of example, an allometric relationship which is well known to mammalogists is the relationship between basal metabolic rate of energy conversion ($\dot{E}_{sm}$, where sm is the standard metabolism) and body mass (M) according to the formulae

$$\dot{E}_{sm\,(Marsupial)} = 2.33\ M^{0.75} \tag{3.2}$$

$$\dot{E}_{sm\,(Eutherian)} = 3.44\ M^{0.75} \tag{3.3}$$

Thus a marsupial of given body size tends to have an $\dot{E}_{sm}$ which is about 70% of that of a comparable eutherian.

Certain species differ substantially from this relationship so there is some overlap between the metabolic rates of the two groups. However, no marsupial has a metabolic rate equivalent to or greater than that predicted by the eutherian relationship (the highest reliable estimate is that for *Chironectes minimus*, with an $\dot{E}_{sm}$ which is 95% of that predicted for eutherians (Fleming, 1982)). By contrast, several eutherians have an $\dot{E}_{sm}$ which is less than the minimum recorded among marsupials (i.e. *Lasiorhinus latifrons* has an $\dot{E}_{sm}$ which is 42% of that predicted for eutherians). We may therefore conclude that in contrast to eutherians, marsupials have a low and comparatively invariant $\dot{E}_{sm}$. The physiological consequences of this difference have been discussed elsewhere (Hume, 1982), and we will return to the ecological significance of this difference later in this chapter.

As a final caution, we should point out that allometric equations, like other regressions, apply only to the range of data from which they have been calculated (Gould, 1971), yet they are usually written as if they applied to an infinite range of x. This means that extrapolation from a sample of limited body size range is unwise, particularly as closely-related taxa may have a shallower slope (k) than distantly-related taxa (Lande, 1979) (Fig. 3.1).

Despite these limitations, the allometric procedure is a powerful comparative tool, allowing us to eliminate the effect of body size in interpretation of adaptive trends while reiterating the importance of body size as an adaptive trait (Western, 1979).

The marsupial/eutherian dichotomy

The small variation in marsupial metabolic rates is a conservative feature associated with low body temperature (see e.g. Hulbert, 1980). By contrast, marsupials are characterised by only a few derived or apomorphic traits. The best-known absolute distinction between marsupials and eutherians concerns the development and architecture of the urogenital ducts (see Tyndale-Biscoe, 1973), but the evolutionary and ecological consequences of these differences are unknown. This is also true of the cranial and neural traits discussed by Johnson, Kirsch & Switzer (1982).

A character of greater potential ecological significance is the different relative maternal investment in gestation and lactation. In eutherians, the duration of gestation (t) is allometrically related to adult weight (M) (Millar, 1981)

$$t = 4.14\,M^{0.24}\ \text{days} \tag{3.4}$$

and correlated with both birth weight and litter weight. These correlations are hardly surprising. By contrast, while marsupials have a variable period of gestation or intrauterine existence (Tyndale-Biscoe & Renfree, in press), the weight of young at birth is independent of gestation length, although

Fig. 3.1. Power functions for brain and body weights in mammals and birds. Closely-related forms have a shallower slope (k) than distantly-related ones. Relative brain size will be overestimated in juvenile specimens. After Lande (1979).

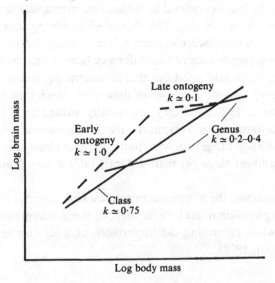

there is a correlation between the weight of a litter (M_{litter}) at birth and maternal body weight (M) (Russell, 1982a):

$$M_{\text{litter (Marsupial)}} = 0.00026\, M^{0.35} \qquad\qquad (3.5)$$

For eutherians the equivalent equation is given by Millar (1981) as

$$M_{\text{litter (Eutherian)}} = 0.11\, M^{0.79} \qquad\qquad (3.6)$$

Combining eqns (3.5) and (3.6) we see that eutherian litters are substantially heavier at all body sizes:

$$M_{\text{litter (Eutherian)}} = 428.7\, M^{0.49}\, M_{\text{litter (Marsupial)}} \qquad\qquad (3.7)$$

Most of the variability in marsupial gestation appears to be confined to the pre-attachment phase of gestation (Fig. 3.2). During this phase the developing embryo makes no contact with the maternal epithelium, and may undergo periods of stasis which are elegantly documented by Selwood (1980) in her study of development in *Antechinus stuartii*. Later intrauterine development, or organogenesis, is highly uniform among marsupials (Fig. 3.2), for reasons which remain uncertain (for discussion see Moors, 1974; Walker & Tyndale-Biscoe, 1978; Rothchild, 1981; Tyndale-Biscoe & Renfree, in press).

Regardless of the cause of this short intrauterine organogenesis, the weight of marsupials at birth is never greater than 1 g. In polytocous species, production of more than one young reduces the weight of neonates

Fig. 3.2. Relative duration of phases of gestation in five different families of marsupials. After Renfree (1980) and Selwood (1980).

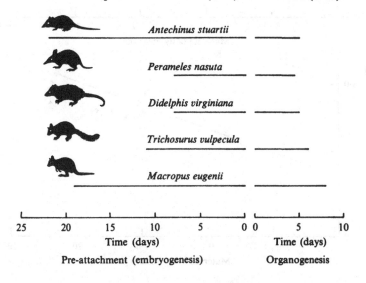

	Antechinus stuartii	
Perameles nasuta	
Didelphis virginiana	
Trichosurus vulpecula	
Macropus eugenii	

25 20 15 10 5 0   0 5 10

Time (days)   Time (days)

Pre-attachment (embryogenesis)   Organogenesis

in the Didelphidae and Dasyuridae, but not among the polytocous Petauridae and Peramelidae (Russell, 1982a) (Fig. 3.3). The latter taxa consequently have a higher relative maternal investment in young at birth than other marsupials, though this parameter has a negative allometric relationship with body size (Fig. 3.4). How this is achieved is uncertain.

Fig. 3.3. Investment in young (mass of neonatus or litter) at birth as a function of maternal body mass. (*a*) Mass of one young; (*b*) mass of the litter. Open triangles, Didelphidae; open circles, Dasyuridae, Myrmecobiidae; filled squares, Peramelidae, Thylacomyidae; open squares, Petauridae; inverted filled triangles, Phalangeridae; filled triangles, Phascolarctidae, Vombatidae; filled circles, Macropodidae. From Russell (1982a), by permission.

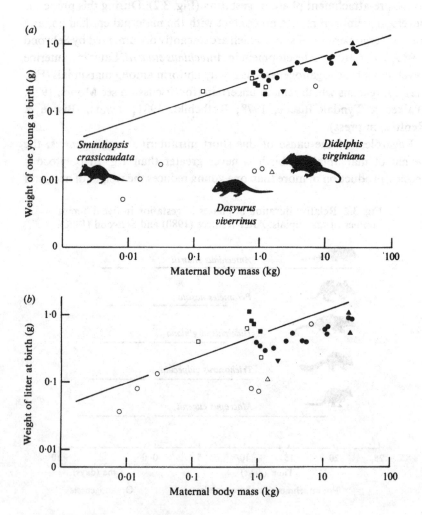

The Peramelidae have more complex placentation than other marsupials, as the allantois vascularises with the chorion to form a placenta similar to, but probably not homologous with the chorioallantoic placenta of eutherians, and the chorionic villi are highly invasive (Padykula & Taylor, 1977; Taylor & Padykula, 1978). By contrast, the choriovitelline placenta of *Pseudocheirus peregrinus*, a petaurid, is not invasive (Sharman, 1961), so a general explanation for this high investment is lacking.

Despite this variation, maternal investment in gestation by marsupials is lower than has been recorded among similarly-sized eutherians. No marsupial litter at birth exceeds 1% of its mother's weight, while among the eutherians considered by Millar (1981) only bears invest an equally low proportion of their body weight in gestation. Some small Insectivora and Rodentia may give birth to litters weighing well over 50% of their mother's weight. However, marsupials are born in a state which is not totally altricial. Neonates combine extremely altricial features with precocial development of features which are necessary to accomplish movement from the birth canal to the teat, and subsequent attachment and suckling (Sharman, 1973). Neonates crawl to the teat using gravitation as a

Fig. 3.4. Relative investment in litter at birth (($M_{\text{litter}}/M_{\text{mother}}$) × 100) as a function of maternal body weight. Open triangle, Didelphidae; open circles, Dasyuridae, Myrmecobiidae; filled squares, Peramelidae, Thylacomyidae; open squares, Petauridae; inverted filled triangles, Phalangeridae; filled triangles, Phascolarctidae, Vombatidae; filled circles, Macropodidae. From Russell (1982a), by permission.

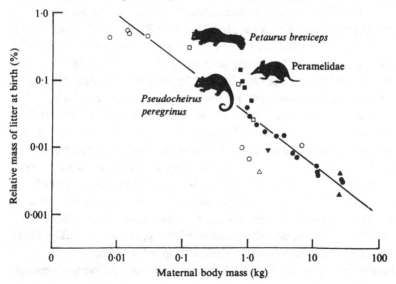

navigational aid (McCrady, 1938; Cannon, Bakker, Bradshaw & McDonald, 1976), and receive little maternal assistance, though in bandicoots they may be anchored to the mother by an allantoic stalk (Stodart, 1977; Lyne & Hollis, 1982). The forelimbs are consequently well developed, as are the shoulder and forelimb muscles. The alimentary tract is complete and functional at birth with the pancreas, gall bladder and villous stomach being particularly well developed (Walker & Rose, 1981). The lungs are partly developed, and there is a functional mesonephric kidney (Buchanan & Fraser, 1918). In contrast to most eutherians, the aortic arches appear simultaneously (Walker & Rose, 1981). Although the eyes, ears and vestibular apparatus are poorly developed, the olfactory bulbs are precocious (McCrady, 1938; Hill & Hill, 1955). Development of the cerebellum, medulla oblongata and the cervical and thoracic regions of the spinal chord is precocious, allowing functioning of the nerves to the forelimbs, suckling mechanisms, lungs and stomach.

We discuss the possible significance of this early and synchronous development later in this chapter, but first comment on what other authors have suggested are the advantages and disadvantages accruing from the small investment in gestation relative to lactation.

The efficiency of lactation

McNab (1978a) has argued that marsupials may be disadvantaged because lactation may be a less efficient means of providing nutrients and energy to the young than is placental exchange. Numerous studies indicate that late lactation is the most energetically demanding period of the life history of eutherians (Brockway, McDonald & Pullar, 1963; Kaczmarski, 1966; Migula, 1969; Millar, 1975, 1978, 1979; Hanwell & Peaker, 1977; Randolph, Randolph, Mattingly & Foster, 1977; Innes & Millar, 1981; Lochmiller, Whelan & Kirkpatrick, 1982; Mattingly & McClure, 1982). This appears to be particularly pronounced in small species (Daly, 1979). Two field studies have provided similar evidence for marsupials. Atramentowicz (1982) showed that lactating female *Caluromys philander* foraged for 2.5 hours longer each night than non-lactating females, a 38% increase. Lee & Nagy (1984) used doubly-labelled water to assess field metabolic rates of *A. swainsonii* (Table 3.1). While other aspects of reproduction affected field metabolism to only a small extent, females during late lactation showed a 76% increase in active metabolic rate. Laboratory studies support this pattern (Fleming, Harder & Wukie, 1981) but are of less relevance because food is supplied *ad libitum*.

These results do not clarify whether there would be a difference in energy expenditure between two females, each having a litter of the same mass, one containing a nursing young and the other being fetal. Lactose synthesis is energetically very demanding (Hanwell & Peaker, 1977), and adaptations for energy conservation are known for some eutherians (Smith & Taylor, 1977). While the energetics and biochemistry of lactation in marsupials is only poorly understood, it is certainly true that the mammary glands of the macropods are the most complex known among mammals (Griffiths, McIntosh & Leckie, 1972; Renfree, 1983). These species show dramatic alteration in the composition of the milk during lactation (see e.g. Green, Newgrain & Merchant, 1980), and different glands may simultaneously produce milk suited to the young of different sizes which feed from them (e.g. Lemon & Bailey, 1966). Whether these adaptations have any consequences for energetics or efficiency is still unknown.

Although these questions are unresolved, they warrant further investigation, particularly as a method is available which should facilitate their resolution (Lee & Nagy, 1984).

Rescindment of reproduction

Several authors who contend that the marsupial mode of reproduction may be viewed in an adaptive framework, or as 'equally fit', have argued that the small investment by marsupials in pregnancy is advantageous in uncertain or variable environments, as termination of reproductive effort can be achieved more easily than in eutherians (Kirsch, 1977a, b; Parker, 1977; Low, 1978). These studies were rightfully castigated by Russell (1982b) and Morton, Recher, Thompson & Braithwaite (1982), who showed that the analyses were based on the biology of the highly

Table 3.1. *Measurements of field metabolic rate and water flux in* Antechinus swainsonii

Sex and condition	Mean body mass (g)	Mean CO_2 production (ml CO_2/g/h)	Water intake (ml H_2O/kg/day)
Juvenile males	32	4.7	564
Juvenile females	26	4.1	572
Mating males	73	3.9	693
Mating females	53	3.8	694
Late lactation females	54	6.6	1337
Post-lactation females	47	4.3	487

After Lee & Nagy (1984).

specialised, large desert-dwelling macropods, and consequently failed to portray adequately the reproductive diversity within marsupials. Russell (1982b) draws attention to theoretical treatments of the tactical decisions involved in abandoning current investment in reproduction (Dawkins & Carlisle, 1976; Boucher, 1977; Maynard Smith, 1977). She places emphasis on the contention of Dawkins and Carlisle (1976) that the value of current reproduction will be related to the proportion this investment represents of total future reproductive prospects. According to this view, the probability of abandonment of a brood should be much higher in a long-lived species which breeds many times (e.g. the macropods discussed by Parker (1977) and Low (1978)) than in a short-lived species which breeds only one or a few times in its lifetime and makes a relatively large energetic contribution to each litter (e.g. *Antechinus* spp., see Cockburn *et al.* (1983)). In these latter species brood reduction is more likely than brood abandonment and indeed occurs widely among the smaller marsupials (e.g. *Sminthopsis crassicaudatus* (Morton, 1978a); *Parantechinus bilarni* (Begg, 1981a); *Phascogale tapoatafa* (Cuttle, 1982a); *Satanellus hallucatus* (Begg, 1981b); *Tarsipes rostratus* (Wooller *et al.*, 1981); *Acrobates pygmaeus* (H. Frey & M. Fleming, personal communication); *Isoodon macrourus* (Gemmell, 1982)). Gemmell's data (1982) are particularly convincing and are illustrated in Fig. 3.5.

It has been contended that *Didelphis virginiana* (Reynolds, 1952),

Fig. 3.5. Reduction in number of pouch young during lactation in *Isoodon macrourus*. Vertical lines are 95% confidence intervals. Data from Gemmell (1982).

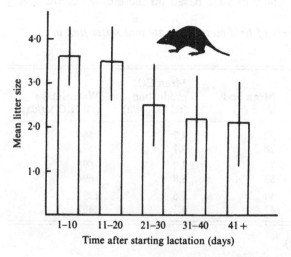

Time after starting lactation (days)

A. stuartii (Marlow, 1961; Settle & Croft, 1982b) and *T. rostratus* (Wooller *et al.*, 1981) will not continue to suckle a litter of one young. However, these results may be an artefact of laboratory conditions, as the first comprehensive field study of weaning success in a polytocous marsupial has shown that *A. stuartii* will suckle to weaning a litter consisting of a single offspring (A. Cockburn & M. P. Scott, unpublished observations).

Although it is crucial to the arguments of Parker (1977), Low (1978) and Kirsch (1977a, b) that the postnatal period is the most common and safest time for brood reduction and cessation of reproductive effort in mammals, they do not attempt to quantify this difference. Eisenberg (1981) has catalogued the possible options for loss of young for females which bear both precocial and altricial young (see Fig. 3.6). Resorption and abortion are available to mothers with precocial young, and are important among species of Lagomorpha, Rodentia and Insectivora (Delany & Happold,

Fig. 3.6. Options for brood reduction or abandonment during gestation or lactation (*a*) for mothers bearing altricial young, (*b*) for mothers bearing precocial young. After Eisenberg (1981).

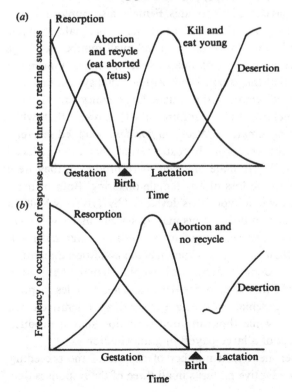

1979; Pelikan, 1981; Morton *et al.*, 1982; Besançon, 1983). The frequency of resorption increases sharply at the end of the breeding season in small rodents and shrews (Besançon, 1983). Infanticide, abortion and other pregnancy interruptions are consequences of social interactions and may have adaptive value in a variety of species (Bruce, 1960; Hrdy, 1979; Boonstra, 1980; Mallory & Brooks, 1980; Labov, 1981; Sherman, 1981). Of greater relevance to this discussion are recent demonstrations that in at least two instances rescindment of reproduction among rodents follows a repeatable pattern, is apparently designed to achieve a similar strategy in each instance, and is amenable to adaptive interpretation. Gosling & Petrie (1981) report that in large postmortem samples of coypus (*Myocastor coypus*) the proportion of females in litters decreases by about 10% around week 14 of the 19-week gestation period. Mothers abort female-biased litters containing few embryos. Embryos from small litters have large amounts of fat, and this is reflected in the large size of offspring at full term and as adults. Mothers conceive postpartum, and as litter size increases throughout the breeding season, the selective abortion enables them to trade a small litter of large, predominantly female embryos for a larger litter that will have smaller neonates. Females also employ another physiological device: they can selectively resorb individual male embryos. Both mechanisms lead to overall production of small numbers of large male neonates and large numbers of small female neonates. Gosling & Petrie (1981) contend that this is consistent with a social system in which males defend groups of females, and where large males are effective competitors. In a different study, McLure (1981) showed that when laboratory-kept lactating females of *Neotoma floridana* were fed severely-restricted diets during this energetically costly period, they often reduced the size of their litters. Where male offspring were produced, they were selectively killed prior to the loss of any female offspring. Both of these examples are consistent with a hypothesis developed by Trivers & Willard (1973), who argued that in a polygynous mating system, males will have more variable reproductive success than females, because fewer males will gain access to females than vice versa. Males in better condition than other males will consequently obtain greater advantage than females having the same advantage of condition over other females. Thus females in poor condition should produce female offspring, each with a moderate chance of reproductive success, while those in good condition should produce males, with the prospect of a large payoff in grandchildren.

These results are pertinent to a number of aspects of the preceding discussion. They show adaptive patterns in all three of the responses that eutherians might use to rescind reproduction and lend support to the idea that litter reduction in eutherians is of broad occurrence and importance. This leads us to agree with Russell (1982b) and Morton *et al.* (1982) that selection for ease of desertion in the face of environmental uncertainty is unlikely to have been an important factor in shaping the differences between the marsupial and eutherian modes of reproduction.

Reproductive rate

Parker (1977) argues that marsupials invest fewer resources than do eutherians during the course of reproduction as a consequence of slower development. This argument ignores the question of the relative efficiency of lactation and gestation, but provides a convenient starting-point for the comparison of marsupial and eutherian reproductive rates. Effective comparison requires examination of a number of life history traits, including litter size, parental investment in litters and individual young, and growth.

Litter size.

Several authors have proposed that litter size (as distinct from litter weight (see eqns (3.5), (3.6), (3.7)) may be expressed as an allometric function of body mass (Millar, 1977, 1981; Tuomi, 1980), and relationships of this sort have been used to derive values for other reproductive parameters (Millar, 1981). Plots of the litter size/body mass relationship suggest that reductionist equations are inappropriate in this instance (Fig. 3.7). Regression statistics are likely to be drastically affected by the tendency of large mammals to produce a single young. Regression is of limited value in instances where for a large proportion of the range of the independent variable (body mass) the dependent variable (litter size) is constrained to a minimum value (one) (Austin, 1971). Further, some small mammal groups have a litter size which increases with body mass (e.g. tenrecs; Eisenberg, 1981).

The data presented in Fig. 3.7 illustrates a number of points. Carnivorous marsupials have larger litters than similarly-sized herbivores. The polygons for eutherians and marsupials are rather similar, though statistical comparison is precluded because Eisenberg (1981) only presents extreme values. This leads us to conclude that on the basis of available data, marsupial litter size reflects similar evolutionary pressures to those operating on eutherians. More precise analysis is also limited because many eutherian species lack marsupial analogues of similar body mass (Chapter 2). We discuss intraspecific variation in litter size at greater length in Chapter 6.

Maternal investment.

The investment by mothers in individual offspring at weaning in marsupials and eutherians is remarkably similar (Table 3.2), leading Russell (1982a) to suggest that constraints are similar in both groups. The relative investment in litters at weaning (PI = mass of litter at weaning × 100/maternal body mass), is allometrically related to body mass (*M*) in marsupials (Russell 1982a) (Fig. 3.8)

$$PI_{\text{Marsupial}} = 78.6 \, M^{-0.29} \tag{3.8}$$

with investment by bandicoots at weaning being rather low. For comparable eutherians, we can calculate (see Fig. 3.8)

$$PI_{\text{Eutherian}} = 66.4 \, M^{-0.27} \tag{3.9}$$

and these equations are not significantly different.

Fig. 3.7. Litter size in marsupials and eutherians of different maternal body mass. Open triangles, Didelphidae; open circles, Dasyuridae, Myrmecobiidae; filled squares, Peramelidae, Thylacomyidae; inverted open triangles, Tarsipedidae, Burramyidae; open squares, Petauridae; inverted filled triangle, Phalangeridae; filled triangles, Phascolarctidae, Vombatidae; filled circles, Macropodidae. Eutherian data from Eisenberg (1981). Marsupial data from Chapters 4 and 5 in this volume.

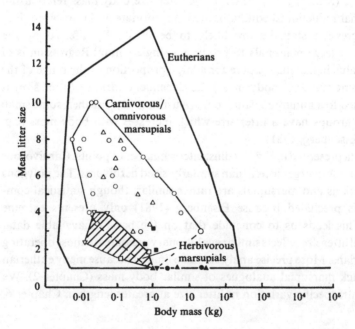

Table 3.2. *Means and ranges of investment by mothers in individual offspring at weaning in mammals*

	Size of individual (% of maternal weight)	Source
Marsupials	35 (9–60)	Russell (1982a)
Eutherians	37 (10–65)	Millar (1977)

Fig. 3.8. Relative investment in litters at weaning (*PI*) as a function of maternal body mass in eutherians and marsupials. Symbols for marsupials: open triangles, Didelphidae; open circles, Dasyuridae, Myrmecobiidae; filled squares, Peramelidae, Thylacomyidae; inverted open triangles, Tarsipedidae, Burramyidae; open squares, Petauridae; inverted filled triangles, Phalangeridae; filled triangles, Phascolarctidae, Vombatidae; filled circles, Macropodidae. Marsupial data from Russell (1982a), by permission. Eutherian data from the following sources. For Insectivora (crosses), *Cryptotis parva*, Conaway (1958); *Suncus murinus*, Dryden (1968); *Condylura cristata*, Petersen & Yates (1980); For Macroscelidea (addition sign), *Rhynchocyon chrysopygus*, Rathbun (1979); For Tenrecomorpha (circle with a dot), *Echinops telfairi*, Gould & Eisenberg (1966); For Scandentia (inverted U), *Tupaia belangeri*, Martin (1968); For Carnivora (inverted Vs) *Mustela nivalis*, Heidt, Petersen & Kirkland (1968); *Suricatta suricatta*, Hinton & Dunn (1967); *Vulpes vulpes*, McIntosh (1963); *Procyon lotor*, Stuewer (1943); *Uncia uncia*, Eisenberg & Hemmer (1972); For Artiodactyla (square with a dot), *Odocoileus hemionus*, Millar (1977); For Rodentia (diamonds), *Mystromys albicaudus, Neotoma albigula, Nyctomys sumichrasti, Peromyscus leucopus, Sigmodon hispidus, Meriones hurricanae, Tatera brantsii, Arvicola terrestris, Clethrionomys glareolus, Ondatra zibethica, Dipodomys deserti, Liomys pictus, Perognathus californicus, Mesembrionys gouldi, Geomys pinetis, Jaculus jaculus*, Millar (1977); For Lagomorpha (triangles with a dot), *Ochotona princeps, Sylvilagus aquaticus*, Millar (1977).

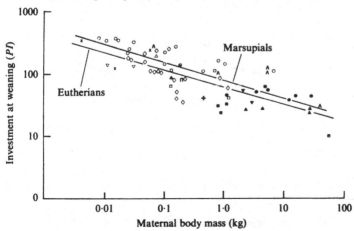

The period required to produce a litter may influence both the temporal spread of reproductive effort and limit fecundity, particularly when the availability of resources is unpredictable or constrained seasonally. Although many young mammals continue to associate with their mothers after weaning, the duration of nutrient transfer to the young (conception to weaning (t_{cw})) is a useful measure of the duration of maternal investment (Braithwaite & Lee, 1979). Russell (1982a) criticised the use of this parameter because of the variability in the preimplantation phase of gestation. We retain it in this analysis both because this variability may have ecological importance, and because the measure also permits comparisons between marsupial and eutherian species.

Fig. 3.9. Time from conception to weaning (t_{cw}) as a function of maternal body mass in carnivorous/omnivorous eutherians and marsupials. Symbols as in Fig. 3.8. Marsupial data from Russell (1982a), Charles-Dominique *et al.* (1981), Cuttle (1982a) and Fanning (1982). Eutherian data from Ewer (1973) and Eisenberg (1981). Eutherian species included (in increasing body mass) are *Cryptotis parva, Sorex araneus, Crocidura russula, Neomys fodiens, Elephantulus myurus, Mustela nivalis, Talpa europea, Tupaia belangeri, Hemicentetes semispinosus, Urogale everetti, Herpestes aropunctatus, Cynictis penicillata, Suricatta suricatta, Erinaceus europaeus, Galidia elegans, Fennecus zerda, Vulpes velox, Felis catus, Vulpes vulpes, Procyon lotor, Lutra lutra, Nyctereutes procyonoides.* Marsupial species are *Planigale gilesi, Ningaui yvonnae, P. maculatus, Sminthopsis crassicaudata, Antechinus stuartii, Pseudantechinus macdonnellensis, A. swainsonii, Marmosa robinsoni, Parantechinus bilarni, Dasycercus cristicauda Dasyuroides byrnei, Phascogale tapoatafa, Caluromys philander, Philander opossum, Dasyurus viverrinus, Perameles nasuta, Isoodon macrourus, Bettongia penicillata, Didelphis virginiana, Dasyurus maculatus, Hypsiprymnodon moschatus, Sarcophilus harrisii.*

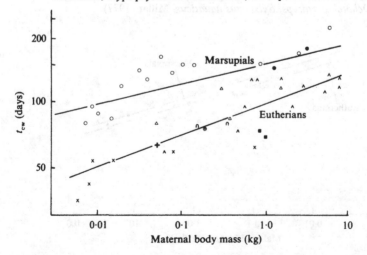

Braithwaite and Lee (1979) have argued that small marsupials take longer to produce a litter than similarly-sized eutherians. This contention was not supported statistically, and the comparison is obscured for larger animals by the inclusion of Carnivora with delayed implantation and small eutherians which lack dietary analogues of similar body size among the marsupials (see Chapter 2). For example, small marsupials are mainly insectivorous, omnivorous or nectarivorous. There are no non-volant nectarivorous eutherians, but many small herbivores. In addition, herbivorous marsupials and eutherians have longer periods of dependence than similarly-sized carnivorous species. To circumvent these difficulties, we have restricted our comparison to the carnivorous/omnivorous marsupials and their eutherian counterparts which lack delayed implantation, as Ewer (1973) argues convincingly that delayed implantation functions to reduce the time that some carnivores take to produce and raise the first litter of the breeding season. Among eutherians from diverse lineages there is a strong allometric relationship between time from conception to weaning (t_{cw}) and body mass (M) (Fig. 3.9):

$$t_{cw\,(Eutherian)} = 99.7\,M^{0.17}\ \text{days} \tag{3.10}$$

Among marsupials, the Peramelidae have very short periods of dependence, which reflects short lactation (Fig. 3.10), short gestation, small weight at

Fig. 3.10. Duration of lactation as a function of maternal body weight in marsupials. Open triangles, Didelphidae; open circles, Dasyuridae, Myrmecobiidae; filled squares, Peramelidae, Thylacomyidae; inverted open triangles, Tarsipedidae, Burramyidae; open squares, Petauridae; inverted filled triangles, Phalangeridae; filled triangles, Phascolarctidae, Vombatidae; filled circles, Macropodidae. From Russell (1982a), by permission.

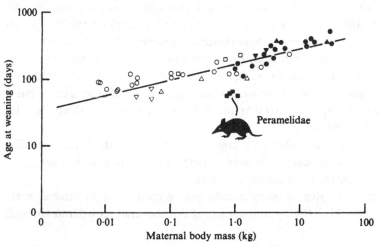

weaning (Russell, 1982a) and rapid growth (see below). If we exclude bandicoots, we can calculate:

$$t_{cw\,(Marsupial)} = 151.4\,M^{0.093}\;\text{days} \qquad (3.11)$$

Thus eutherians and marsupials (excluding bandicoots) can be compared by combining eqns (3.10) and (3.11):

$$t_{cw\,(Marsupial)} = 1.52\,M^{-0.076}\,t_{cw\,(Eutherian)} \qquad (3.12)$$

confirming that the difference between the two taxa is most pronounced for small species.

Growth.

Life history processes encompass more than reproductive behaviour, yet a life history theory of ontogenetic processes is only poorly developed. Ontogeny includes several dissociable elements (Gould, 1977):

(1) Growth or increase in spatial dimensions and weight.

(2) Differentiation or change in shape, complexity and organisation.

(3) Maturation and acquisition of function.

Growth may be affected, but not necessarily limited by both the rate at which resources are made available and the efficiency with which they are converted to tissues. For example, heterotherms have a higher growth efficiency than homeotherms, and carnivores have a higher efficiency than herbivores(Schroeder, 1981). However, endothermy means that metabolism can occur at a high and constant rate irrespective of environmental disturbances (Calow & Townsend, 1981).

Interspecific differences in growth are often interpreted as the result of trade-offs between selective factors which favour longer or shorter periods of growth. For example, Lack (1968) suggested that postnatal growth of birds was a compromise between food availability, which would limit the energy allocated to growth, and patterns of chick mortality. Where predation was intense or hypothermia a problem, fast growth would be favoured. Similar arguments have been applied to eutherians (Case, 1978; Robbins & Robbins, 1979). In contrast to this viewpoint, other authors have argued that metabolic limitations to growth are of limited importance. Ricklefs (1979) points out that internal constraints on growth in birds may operate at two levels:

(1) The availability of energy and nutrients to the chick, including their allocation between growth, reproduction and maintenance (organism or metabolic level).

(2) The rates of proliferation and growth of cells (tissue level). Ricklefs (1979) provides strong evidence that growth rates in birds

with precocial offspring are determined at the tissue level by the growth of the slowest-growing tissue, the skeletal muscle of the legs. Sacher & Staffeldt (1974) used an analogous argument to suggest that the pacemaker of prenatal growth in eutherians is the brain.

Investigation of these hypotheses is plagued by methodological confusion. Case (1978) described growth in eutherians by measuring increase in weight (g/day) over the relatively linear phase of growth from the time when 5% of growth had been completed until the time when 30–50% of growth had been completed. For a small sample of marsupials he used the period beginning midway through pouch life and ending when 30–50% of growth had been completed. He concluded that marsupials grow slowly, at similar rates to anthropoid primates. Russell (1982a) obtained a different result for marsupial growth when she measured growth (g/day) over the period from final exit from the pouch until weaning. Maynes (1976) had earlier argued that this was a period of relatively linear growth in *Macropus parma*. These analyses are of dubious significance for a number of reasons. First, Russell's measure spans from only 5% of growth in some species (e.g. *Didelphis* or *Isoodon*) to about 50% of growth in others (e.g. *Ningaui* sp.). This limits the potential for interspecific comparisons. Second, allometric variation in grams gained per day is probably of rather trivial significance. That a large whale can grow at 66 kg/day while a small shrew cannot, does not really have any ecological (or other?) significance. Third, these studies assume linearity, while mammalian growth usually takes a sigmoid form, with slow initial growth followed by a period of rapid increase and then a slackening of growth as adult weight is approached.

Two mathematical descriptions of sigmoid growth are available, but neither has been applied to marsupial growth. The logistic curve is widely used in analyses of population growth, and is symmetrical about the inflection point, or time of maximum growth. The Gompertz curve probably describes mammalian growth more accurately (Laird, 1966a, b), and is asymmetrical, approaching the asymptote more gradually than would a logistic curve with a similar early trajectory (Fig. 3.11). The Gompertz growth equation has the form

$$dW/dt = -\alpha W(\ln W) \qquad (3.13)$$

where dW/dt is the absolute growth rate, W is the fraction of the asymptotic weight attained, and α is a constant that is directly proportional to the rate of growth. This constant is the rate of decay of growth, and is particularly suitable for interspecific comparison, as it is independent of the degree of precociality at birth. This is useful in marsupial/eutherian comparisons, and also for analysis of interspecific variation between marsupial families, as the Dasyuridae and Didelphidae are both born at a very small size (Fig. 3.3). Kaufmann (1981) has recently provided a useful technique for deriving α, which should make the use of Gompertz equations accessible to scientists with only limited mathematical skills.

We have fitted Gompertz equations to all of the published marsupial growth data for which birth weight and sigmoid growth curves are available (Fig. 3.12). The rate of decay of growth varies allometrically with asymptotic body mass (Peramelidae are excluded for consistency):

$$\alpha_{\text{Marsupial}} = 0.014\,M^{-0.17} \tag{3.14}$$

Small species thus grow relatively faster than large species, which relates to the speed with which they attain sexual maturity. The different results of Case (1978) and Russell (1982a) reflect their measurement of absolute growth over a short segment of the total duration of growth. Inspection of the scatter of points in Fig. 3.12 shows that seven species have slow growth for their body size, and each of these falls exactly on a straight line. It is tempting to hypothesise that these species grow at a basal rate for marsupials. Five other species exhibit faster growth, and we argue in Chapters 4 and 5 that each of these, except *Thylogale billardierii*, has undergone selection for rapid maturity in contrast to their close relatives. We might therefore predict that a species such as *Trichosurus caninus*, in which females mature in their third and fourth years, should be much closer to the basal rate of growth than *T. vulpecula*, in which females mature in their second year (How, 1978). There are some data supporting this proposition (How, 1976), but testing of these hypotheses awaits publication of sigmoid growth curves for *T. caninus* and many other species. It is also worthy of note that *Thylogale billardierii* occurs in Tasmania, where the climate is colder and the plant-growing season is shorter than in the habitats of the other marsupials for which we have data. This may have

Fig. 3.11. The Gompertz and logistic curves for growth.

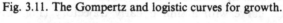

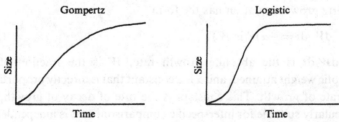

led to selection for rapid growth. All of these comparisons need to be viewed in the light of intraspecific variation in growth (Parker, 1977).

Our results also facilitate comparison with eutherians and birds. Ricklefs (1973, 1979) collated growth curves for 148 species of birds, to which he

Fig. 3.12. The Gompertz constant (α) as a function of asymptotic body mass in marsupials and eutherians. Open circles, Dasyuridae, Myrmecobiidae; filled square, Peramelidae, Thylacomyidae; inverted open triangle, Tarsipedidae, Burramyidae; open square, Petauridae; inverted filled triangle, Phalangeridae; filled circles, Macropodidae. Original curves for marsupials from the following sources (in order of increasing body mass): *Ningaui yvonnae*, Fanning (1982); *Sminthopsis murina*, Fox & Whitford (1982); *Antechinus stuartii*, Marlow (1961); *Cercartetus caudatus*, Atherton & Haffenden (1982); *Petaurus breviceps*, Smith (1979); *Perameles nasuta*, Lyne (1964); *Peradorcas concinna*, G. D. Sanson & P. Fell (personal communication); *Bettongia lesueur*, Tyndale-Biscoe (1968); *Trichosurus vulpecula*, Lyne & Verhagen (1957); *Macropus parma*, Maynes (1976); *Thylogale billardierii*, Rose & McCartney (1982); *M. giganteus*, Poole, Carpenter & Wood (1982). Eutherian curve calculated from Laird (1966a).

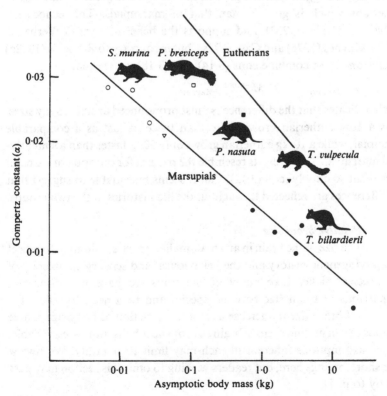

fitted the logistic growth equation, and showed that the constant of decay (β) in the logistic equation could be described thus:

$$\beta = 0.11\,M^{-0.34} \tag{3.15}$$

Ricklefs (1973) showed that at the inflection point

$$\alpha = 0.68\,\beta \tag{3.16}$$

By combining eqns (3.15) and (3.16) we can argue that at the time of maximum growth

$$\alpha_{\mathrm{Bird}} = 0.071\,M^{-0.34} \tag{3.17}$$

Appropriate transformations have not been performed for growth curves of many species of eutherians, and such a task would be too exhaustive to attempt here. Analysis of Laird's (1966b) original use of Gompertz curves to describe growth rates of a variety of eutherians, some of which were domesticated and consequently of dubious generality, reveals:

$$\alpha_{\mathrm{Eutherian}} = 0.025\,M^{-0.23} \tag{3.18}$$

These results suggest that the growth rate of birds is greater than that of eutherians which is greater than that of marsupials. This echoes the analysis of Calder (1982), and supports the belief of Lyne & Verhagen (1957), Maynes (1976) and Case (1978), but is contrary to Russell's (1982a) conclusion. If we combine eqns (3.14) and (3.18) we see that

$$\alpha_{\mathrm{Eutherian}} = 1.76\,M^{-0.061}\,\alpha_{\mathrm{Marsupial}} \tag{3.19}$$

which indicates that the difference is most pronounced at small body sizes. Thus a 10-g eutherian grows more than twice as fast as a comparable marsupial, while a 10-kg eutherian grows only 50% faster than a similarly-sized marsupial. These results resemble the pattern for duration of parental investment so closely (eqn (3.12)) that it seems reasonable to suggest that the differences are reflected throughout the life histories of the two groups.

Relative brain size

The size of the brain in any mammalian species is thought to reflect both phylogenetic history and the behavioural and sensory complexity of that species' niche. Like many of the traits we have just discussed, comparisons of brain size between species and taxa need to reflect the tendency of brain size to increase as a power function of body size. Three methods of calculating relative brain size, or encephalisation, are available, but the assumptions inherent in each may limit their utility. We review these shortcomings here, but readers wishing to omit this section may pass directly to p. 71.

(1) In the first method, encephalisation is expressed as a fraction of the value predicted by a power function relating brain size to body size across the full range of mammalian body sizes (see e.g. Jerison, 1973; Eisenberg, 1981; Hofman, 1982a, b). Although original calculations were based on a function with an exponent of 0.67 (Jerison, 1973), and it has been widely accepted that brain size is scaled to keep pace with changes in body surface area, recent studies have agreed that the true exponent is somewhat higher (Table 3.3). We have already pointed out that this relationship does not apply to subordinate taxonomic groupings or during ontogeny (Fig. 3.1). The equivalent relationships for several eutherian orders are given in Table 3.3, and with the exception of the Chiroptera, the exponents for each order are lower than that for mammals as a whole (0.75). Consequently, large species will appear to have low encephalisation relative to small species, even though the scaling relationship within a taxon is constant. Thus interpretation will be confounded in any analysis where parameters of interest are themselves size-dependent (e.g. diet-associated). This problem is also apparent in marsupial/ eutherian comparisons (see Table 3.3).

(2) The second technique attempts to circumvent this difficulty by selecting a reference group from which comparisons may be

Table 3.3. *Power functions relating brain mass to body mass*
($M_{brain} = bM_{body}{}^{K}$)

Taxa	No. of species	b	K	Source
Eutherians	309	0.059	0.76	Martin (1981)
Marsupials and eutherians	547	0.055	0.74	Eisenberg (1981)
Marsupials and eutherians	249	0.064	0.73	Hofman (1982a)
Marsupials (Australian)	89	—	0.65	Nelson & Stephan (1982)
Marsupials (American)	17	0.094	0.61	Eisenberg & Wilson (1981)
Insectivora	—	0.055	0.68	Eisenberg (1981)
Chiroptera	—	0.046	0.79	Eisenberg (1981)
Primates	—	0.023	0.67	Eisenberg (1981)
Rodentia	—	0.102	0.63	Eisenberg (1981)
Cetacea	—	9.120	0.23	Eisenberg (1981)
Carnivora (excluding Pinnipedia)	—	0.040	0.57	Eisenberg (1981)
Artiodactyla	—	0.039	0.54	Eisenberg (1981)

derived in the same way as for the preceding technique. Thus Stephan, Nelson & Frahm (1981) used the Vespertilionidae (excluding *Tylonycteris*) as the reference group for examining brain size variability in Chiroptera, and Nelson & Stephan (1982) used the Dasyuridae (excluding *Planigale*) as a reference group for the Australian marsupials. In both studies a phylogenetically-heterogeneous group they called the 'basal Insectivora' was used as a reference point for the analysis of brain enlargement. These reference groups share the characteristic of having the smallest relative brain size among the subgroups for which large samples are available. We feel that this technique offers no real advantages unless the allometric exponents for each subgroup are identical, which is not true for any of the studies cited. There is an additional disadvantage. The technique involves extrapolation rather than interpolation, and this is invalid for regression analysis (Gould, 1971; Zar, 1974).

(3) It has been suggested that intrataxonomic variation in the brain size of small mammals and Primates can be interpreted by calculating separate slopes for families and calculating values for each genus as deviations from the line of best fit (comparative brain size) (Clutton-Brock & Harvey, 1980; Mace, Harvey & Clutton-Brock, 1981). This procedure enabled Clutton-Brock and his co-workers to comment on the effect of habitat complexity, diet and activity timing on relative brain size. While their analyses are probably the most statistically sound procedures currently in use, they also suffer from a number of problems. First, even more than the two techniques described above, they assume the functional and evolutionary equivalence of higher taxonomic categories such as genus and family, assumptions about which systematists seldom agree and other biologists seldom care (e.g. Archer & Kirsch, 1977). Second, the degree of deviation for each genus will depend in part upon the range of body sizes it encompasses, unless the scaling for each genus is identical to that for families. Finally, the analysis will obscure the significance of differences at the highest taxonomic level. For example, Mace *et al.*, (1981) suggested that nocturnal small mammal genera have larger relative brain sizes than their diurnal counterparts, even though this difference is not statistically significant. However, the two rodent families with the largest brains are the predominantly diurnal Sciuridae and the exclusively diurnal Caviidae. The mode

of analysis employed ignores the possibility that diurnality and large brain size are evolutionarily interrelated in these two groups.

Despite these difficulties, it is possible to discern a number of trends in the evolution of relative brain size both by comparison of extant forms and by comparison of brain size and form in contemporary mammals with endocasts of extinct species. In a comprehensive treatise on changes in brain size through time, Jerison (1973) argues that many taxa independently exhibit increases in encephalisation. He further argues that the increase in brain size that occurred among ungulates and carnivores on the contiguous continents was a consequence of predator/prey coevolution, where increased brain size in predators led to an increase in brain size in prey, and vice versa. He suggests that this pattern was absent or less evident in the now-extinct neotropical marsupial carnivores and ungulates, and speculates (p. 339):

> a corollary of this hypothesis is that the marsupials could not respond to the kind of feedback of selective pressures that produced the 'relay' in relative brain size in the Holarctic fauna.

Jerison's argument is clearly circular, and his contention concerning the neotropical ungulates is not supported by recent data (Radinsky, 1978, 1981). However, the suggestion does illustrate the widespread belief that marsupial brains are small and conservative in comparison to those of similarly-sized eutherians.

In order to clarify this issue, we have plotted the brain sizes for all the marsupials for which data are available (106 spp.) against body mass and compared this with the convex polygon for all 547 mammals used by Eisenberg (1981) in generating his allometric function relating brain mass to body mass in mammals (Fig. 3.13). The allometric function for brain size against body mass (Table 3.3) and those for a 100% increase and a 50% decrease in brain weight are also plotted. Small marsupials (< 60 g) have a brain size close to the mammalian mean (*Burramys* and *Planigale* are small-brained exceptions). Other species typically have brain sizes lower than the mean, though this is not true of the Potoroinae, the exudate-feeding/insectivorous Petauridae, and *Myrmecobius*. Small brain size is most pronounced in the folivorous Petauridae, and all large marsupials (> 5 kg). At sizes greater than 10 kg, marsupials fall at the bottom limit of mammalian brain sizes. This would be even more pronounced if tenrecine insectivores were excluded (Eisenberg, 1981).

The data are also suggestive of some environmental influence on brain size. For example, folivorous possums have a much smaller brain size than exudate-feeding/insectivorous species, regardless of the criterion of comparison (see also Nelson & Stephan, 1982). A similar pattern occurs in both Primates (Clutton-Brock & Harvey, 1980), and Rodentia (Mace *et al.*, 1981). The uniformity of this result suggests that a useful general model of the importance of increased or decreased encephalisation may be possible. A corollary of this argument is that increased brain size does have functional significance. For example, it is interesting that the exudate-feeding/insectivorous Petauridae appear to have the most complex social organisations known among the marsupials (Chapter 5). Other environmental factors thought to influence the evolution of brain size include habitat complexity (see e.g. Eisenberg & Wilson, 1981), diel activity (see e.g. Charles-Dominique, 1975) and interspecific interactions (Mace & Eisenberg, 1982).

Changes in the relative sizes of the major subdivisions of the brain have

Fig. 3.13. Brain mass as a function of maternal body mass in marsupials. Data from Eisenberg & Wilson (1981) and Nelson & Stephan (1982). Open triangles, Didelphidae; open circles, Dasyuridae, Myrmecobiidae; filled squares, Peramelidae, Thylacomyidae; inverted open triangles, Tarsipedidae, Burramyidae; open squares, Petauridae; inverted filled triangles, Phalangeridae; filled triangles, Phascolarctidae, Vombatidae; filled circles, Macropodidae. The solid regression function is that for all marsupials and eutherians given by Eisenberg (1981); the dotted lines represent doubling and halving curves. Where these are exceeded by the minimum convex polygon for all species, the polygon is plotted. The vertical hatched area encloses many species of Macropodidae.

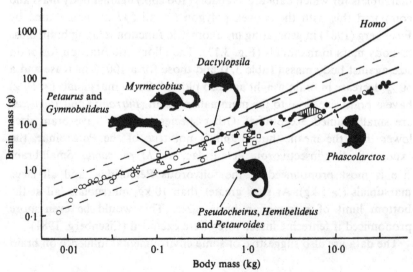

taken place through evolutionary time. These were documented by Jerison (1973) and summarised by Eisenberg (1981, p. 275):

> Early mammals had large olfactory lobes with central projection areas from the first cranial nerve (and)...the seventh cranial nerve. Hearing and the chemical senses were apparently important for the nocturnal early mammals, and hearing requires much central processing, with considerable brain enlargement. Visual information in early mammals did not involve fine form discrimination until they began to invade the arboreal niche. Since much visual information can be processed peripherally, little enlargement of those parts of the brain concerned with vision was discernible in early nocturnal mammals. Increased use of sight and diurnality led to a parallel enlargement in the cortical and subcortical areas receiving neuronal connections from the second cranial nerve and corresponding enlargements in brain mass. The evolution of diurnal habits further accelerated the encephalisation of the Mammalia.

It would be premature to comment on the relative composition of marsupial and eutherian brains (see, however, Meyer, 1981), though a comprehensive project has been undertaken which should resolve this question (J. E. Nelson, personal communication). However, we can comment briefly on the transition from nocturnality to diurnality. It is widely accepted that ancestral eutherians and marsupials were adapted for nocturnal life, and that specialisation for diurnal existence is a derived condition (Taylor's (1980) argument to the contrary is obviously circular). The nocturnal condition is retained in the vast majority of Insectivora, Chiroptera and Dermoptera. Specialist diurnal and nocturnal forms occur sympatrically in the Pholidota, Edentata, Rodentia, Lagomorpha, Macroscelidea, Carnivora, Scandentia, Primates and Artiodactyla. Although many marsupial species are crepuscular (e.g. many macropods) or have an even diel cycle (e.g. *Antechinus minimus*), only one species appears to be exclusively diurnal (*Myrmecobius fasciatus*). We return to the significance of this rarity in the concluding section to this chapter.

Interrelationships between allometric functions
The allometric functions we have identified in the preceding discussion do not imply causality, but have heuristic value in three different ways:
    (1) They facilitate comparison between taxa (e.g. the marsupial/ eutherian dichotomy).

(2) They permit the identification of outliers by taking into account scaling effects (e.g. the rapid ontogeny of bandicoots).

(3) They suggest the existence of functional constraints on reproductive behaviour and physiology associated with body size.

Several authors have sought causal relationships between life history and physiological parameters by comparing the exponents of various power functions. For example, Martin (1981) argues that because basal metabolic rate and brain size in eutherians scale to similar exponents, brain size may be linked to maternal metabolic turnover. This suggestion has several flaws which illustrate the limited utility of attributing significance to similarity between exponents. First, there is no evidence for metabolic limitation to brain growth in eutherians (Sacher & Staffeldt, 1974). Of greater importance, however, is the composite nature of the brain size/body size relationship. Thus while the exponents for metabolic rate approach 0.75 in various mammalian subgroups (including the marsupials), this is not true for brain size (Table 3.3). It therefore appears that Martin (1981) is comparing a fundamental scaling relationship with one which is at least in part an epiphenomenon.

McNab (1978a, b, 1979, 1980a, b, 1982, 1983) has suggested that differences in fecundity in mammals reflect variability in $\dot{E}_{sm}$, with animals with a high relative metabolic rate possessing a greater intrinsic rate of natural increase r_m (where m stands for Malthusian). This argument appears to run counter to the widely accepted notion that the energy which organisms devote to reproduction will be restricted by the increased cost of growth and maintenance (see e.g. Calow, 1979). McNab (1980a) circumvented this difficulty by suggesting that many mammals reproduce during a season when food is not limiting. Thus the factor limiting growth and reproduction might be the rate at which energy can be used, and as a high body temperature and rate of metabolism will mean a higher rate of biosynthesis, a faster growth rate, a shorter generation time and a higher r_m will ensue. McNab's arguments are based for the most part on the comparison of exponents, and it is true that r_m scales with body mass in a similar manner to $\dot{E}_{sm}$ in a variety of organisms (see e.g. Fenchel, 1974; Blueweiss et al., 1978). However, such relationships do not imply causality. Hennemann (1983) used the procedure of expressing r_m and $\dot{E}_{sm}$ for each species as a percentage of that predicted by the power function relating each parameter to body mass. Although there was a statistically significant relationship between these deviations, there was a very high level of residual variance, and numerous qualifications were necessary to permit interpretation of the correlation.

A more comprehensive analysis of similarity in allometric exponents was attempted by Lindstedt & Calder (1981). They suggest the existence of a 'pace of life' in mammalian and avian biological phenomena which scales in proportion to $M^{0.25}$. They present scaling functions for a variety of physiological and life history traits, and contend (pp. 3–4):

> The basic maintenance of an animal's internal composition is a logistical problem of obtaining the right amounts (in regard to body mass) at the right times (in regard to mass-dependent time scales), for all of the many simultaneous and interrelated functions of the living state... These properties may have set the physiological time scales of birds and mammals in proportion to $M^{0.25}$. In fact, most cyclical events do transpire on time scales proportional to $M^{0.25}$. Consequently, it may not be necessary to invoke special meanings or unique explanations to describe how one rate or another scales to body mass... Hence, we propose no specific evolutionary explanation for the scaling of physiological time.

Relations of this sort will undoubtedly continue to be discovered but we believe their importance may have been overstated. We have described two measures of ecologically significant rate processes for marsupials, the pace of growth (α), and the duration of parental investment (t_{cw}). Exponents for these processes differ from both 0.25 and from the equivalent relationship for eutherians. As other relationships, e.g. $\dot{E}_{sm}$ and PI at weaning, scale closely to 0.25 in both marsupials and eutherians, it appears that the pace of life in the two groups may be fundamentally different and uncoupled from $\dot{E}_{sm}$ in marsupials (a similar conclusion is alluded to by McNab (1982)). In their analysis of home range patterns in birds and rodents, Mace & Harvey (1983) comment that the 0.25 exponent for metabolic rate is derived for post-absorptive and resting animals, and may be irrelevant to the metabolic needs and processes of active animals. Indeed, the biochemical and physiological bases of active and resting metabolism are fundamentally different (Taigen, 1983). The energetic requirements of active animals are much less well known than the basal requirements, though research in progress by K. A. Nagy (personal communication) will help to remedy this deficiency. One published comparison of the cost of activity is Baudinette's (1980) analysis of locomotion in dasyurids. He showed that

$$\dot{E}_{run\,(Eutherian)} = 1.78\,M^{-0.06}\,\dot{E}_{run\,(Marsupial)} \tag{3.20}$$

where $\dot{E}_{run}$ is the energy expended in locomotion in ml $O_2/g/km$. Although Baudinette (1982) has subsequently expressed doubt as to the significance of this result, we feel it is worth drawing attention to the similarity of this relationship to the values we have derived in eqns (3.12) and (3.19). These functions provide preliminary evidence that the pace of life is slower in marsupials than in eutherians and also differs in scaling, so that the difference is most pronounced for small species.

The dichotomy: a summary

The difference between the eutherian and marsupial radiations that we have established in this chapter and in Chapter 2 are summarised in Table 3.4, together with apparent exceptions to our idealised notions of 'marsupialness' and 'eutherianness'. Among all the characters we have considered, only the relative investment at birth distinguishes marsupials absolutely from eutherians. Even the brown bear (*Ursus arctos*), in which the litter at birth is only 0.6% of maternal body mass (Millar, 1981), still exceeds the value predicted for a similarly-sized marsupial by two orders of magnitude.

Other differences are only of degree, and reflect five major sets of characters where the marsupial radiation is restricted in comparison to that of the eutherians:

(1) The conservative and generally low metabolic parameters.
(2) The slower pace of growth and rate of reproductive investment, which may imply a lower cost of reproduction per unit time. This difference is most pronounced for small species, and the bandicoots represent an apparent exception.
(3) The conservative brain size variation, and the small brains of the larger marsupials.
(4) The conservative variation in body size among marsupials occupying major feeding niches.
(5) The rarity of fossorial, aquatic and specialist diurnal foragers. While other authors have drawn attention to the absence of flying marsupials (e.g. Lillegraven, 1975), flight has evolved only once in the eutherians, while fossoriality, aquatic specialisation and diurnality have evolved repeatedly. Consequently we attribute greater significance to the rarity of these modes.

The conservatism of the marsupial radiation

We have presented quantitative evidence that the ecological opportunities exploited by marsupials are restricted in contrast to those of eutherians. There are several possible bases for this conservatism:

(1) Selection on the marsupials has not required the acquisition of the

Table 3.4. *The differences between marsupials and eutherians*

Trait	Nature of difference in marsupials	Size range in which most pronounced	Marsupial exceptions	Eutherian exceptions
Maternal investment				
Litter size	No herbivores with large litters	Small species	—	—
Investment at birth	Much lower	Large species	—	? Ursidae
Investment at weaning	Very little	—	Low in Peramelidae	Not quantified
Duration of investment	Much longer	Small carnivores	Peramelidae, some Didelphidae	—
Postnatal growth	Much slower	Small species	Peramelidae	Not quantified
Metabolism				
Basal metabolic rate	70% of eutherian value, more conservative	All sizes	*Chironectes*	See McNab (1980a)
Body temperature	Lower, more conservative	All sizes	—	See McNab (1980a)
Thermal conductance	Very little	—	—	—
Energetics of locomotion	Lower	Small species	Not quantified	Not quantified
Relative brain size	Lower, more conservative	Large species	Exudate feeder/ insectivores, *Myrmecobius*, Potoroinae	Tenrecine insectivores, *Solenodon*
Body size				
Carnivore/omnivores	Invariant	Large species	—	—
Herbivores	Invariant	All sizes	? Diprotodontidae	—
Granivores	Invariant	Only one extant species	? Argyrolagidae	—
Exudate feeder/ insectivores	—	—	—	—
Diurnality	Very rare	Large species	*Myrmecobius*, ? *Hypsiprymnodon*	—
Foraging arenas				
Aboreal	Common in both groups	—	—	—
Fossorial	No herbivores	Only one extant species	—	—
Aquatic	Rare	Only one extant species	—	—

more diverse morphology and behaviour exhibited by the eutherians, perhaps as a result of the isolation from the radiation of the eutherians on the contiguous continents, or because the axes of diversification in the two groups are fundamentally different.

(2) Differences are an artefact of the apparently more prolific speciation in the eutherians, so the conservatism in the marsupials reflects small sample size. While this question is presumably amenable to statistical examination, limited understanding of the behaviour of many species may hinder more sophisticated quantitative analysis than we have attempted.

(3) There are biological constraints on the diversification of the marsupials. It is again important to stress that constraint on diversification is fundamentally different from relative efficacy or fitness within a given niche. To illustrate this distinction, Jerison's (1973) discussion of the evolution of encephalisation in mammals (Fig. 3.14) showed not just an increase in relative brain size through time, but also a diversification of brain sizes. Unlike an idealised representation of directional selection (Fig. 3.14), the development of large-brained phenotypes has not led to the

Fig. 3.14. Evolutionary shifts in phenotype frequencies in species and higher taxa. (*a*) A conventional model of directional selection, applicable to species. Arrow denotes direction of selection. (*b*) Diversification of brain size in ungulates through time. After Jerison (1973). Note that small-brained forms are still represented in Recent faunas.

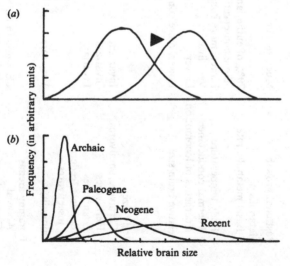

extinction of small-brained phenotypes. There are apparently still niches for which small relative brain sizes are adequate.

Determination of the relative importance of these influences is obviously fraught with difficulty. In attempting to shed some light on this issue, we examine evolution in one character state, the specialisation for diurnal foraging. We have previously commented that Jerison (1973) and Eisenberg (1975, 1981) felt that expansion into diurnal niches was one of the stimuli for acceleration of encephalisation in mammals. Charles-Dominique (1975) has elaborated on these ideas and proposed that specialisation for diurnality is also accompanied by acquisition of a number of correlated characteristics which include large body size relative to sympatric nocturnal forms, greater behavioural complexity and tendency towards gregariousness. Social communications are made principally by vocal and visual signals, and colour vision is frequently developed. The relevance of these changes to the preceding discussion is obvious.

Charles-Dominique's (1975) contentions are supported by several observations, including the high relative brain size of *Myrmecobius fasciatus*, the only diurnal marsupial. In addition, nocturnal species often retain numerous primitive morphological and behavioural characteristics. For example, the most morphologically primitive extant Artiodactyla are the mouse-deer, or Tragulidae (Eisenberg, 1981). All species are small, nocturnal, do not form cohesive social structures and rely heavily on olfaction (Dubost, 1975; Ralls, Barasch & Minkowski, 1975). Among the fissiped Carnivora, the nocturnal viverrids retain many conservative morphological features. Waser's (1980) study of the social behaviour of sympatric carnivores on the Serengeti Plains shows clearly that nocturnal species have simple social organisation in contrast to their diurnal counterparts, though he provided a different explanation for the phenomenon. Charles-Dominique's (1975) original study concerned African and neotropical Primates. Indeed, it appears that the transition to diurnality has occurred in twelve of the nineteen major eutherian lineages illustrated in Fig. 3.15. The exceptions (Tenrecomorpha, other Insectivora, Chiroptera and Dermoptera) share with marsupials a variety of primitive characteristics and a small range of body sizes. The ubiquity of the transition, and its occurrence in small geographically-restricted taxa, e.g. the Malagasy lemuriforms, the Scandentia, or even within widely distributed groups like the Artiodactyla, appears to run counter to the argument that differences reflect the more prolific speciation in the eutherians.

Further, this pattern suggests that the absence of the trend in marsupials is not a result of the absence of selection pressure. This is true both because of its frequent occurrence in eutherians and because of the overwhelming evidence we presented earlier that many evolutionary trends are similar in the two groups. This similarity has led, for example, to the extraordinary convergence between the nocturnal prosimians and the Burramyidae and Petauridae (Chapter 2).

Although absence of selection and small sample size are inappropriate explanations in this instance, this does not mean that the role of evolutionary constraint should be accepted by default, particularly as this third hypothesis is most difficult to test. Previous discussions of the marsupial/eutherian dichotomy have identified two possible sources of constraint, chromosome number and heterochrony.

Fig. 3.15. Evolution of diurnal foraging in the main eutherian lineages. Filled stars, diurnal and nocturnal forms occur sympatrically; open stars, diurnal lineage; circles, comparison inapplicable. Dendrogram and systematic placement after Eisenberg (1981).

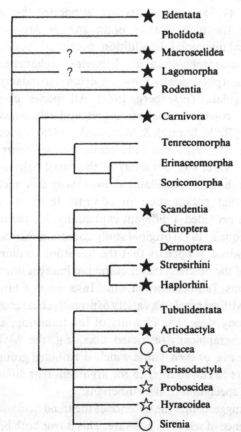

Chromosome number

Lillegraven (1975, p. 719) drew attention to the uniformly low chromosome numbers of marsupials and commented:

> One cannot help wonder...if the generally reduced number of linkage groups (through lowered chromosome numbers) seen in marsupials might not have had some complementary relationship with their lower degree of flexibility in comparison with placentals.

We know of no strong theoretical or empirical evidence relating to this hypothesis and are not convinced it is worth pursuing further. A development of this hypothesis to incorporate recent speculation on the interaction between social structuring of populations and rates of evolution (Wilson, Bush, Case & King, 1975; Bush, Case, Wilson & Patton, 1977) was presented in a later review by Lillegraven (1979). The application of these ideas is restricted by Lillegraven's underestimation of the diversity of social systems in the marsupials, a theme on which we elaborate in Chapter 5.

Heterochrony

A different source of constraint was proposed in an important but widely neglected series of papers by Muller (e.g. 1969, 1973), which were summarised in English by Lillegraven (1975). Muller was a student of Adolph Portmann, whose school had suggested a fundamental distinction between mammals with altricial and precocial offspring. The altricial condition was considered to be primitive (e.g. Portmann, 1939, 1965). Differences between altricial and precocial types were summarised by Martin (1975) (see Table 3.5). These authors consider that the acquisition of precocial offspring was achieved by prolongation of gestation, which in turn facilitated morphological elaboration through two processes:

(1) *phylogenetischer Tragzeitverlangerung* (which Lillegraven translates as phylogenetic extensions), in which the anatomical development at birth is variously refined as a result of the additional time available for ontogeny.

(2) *phylogenetischer Tragzeitdehnung* (phylogenetic expansions), in which increase in development time is not accompanied by increased differentiation of the embryo or organ.

In particular, phylogenetic extensions were thought to permit refinement of structures prior to the acquisition of function at birth. Special attention was paid to brain function and encephalisation (see e.g. Wirz, 1950; Mangold-Wirz, 1966) and led Muller (1969, 1972, 1973) to suggest that as a consequence of the very short intrauterine existence of marsupials, encephalisation would be restricted, and morphological and behavioural diversification would be restricted.

In his review of her research, Lillegraven (1975) disregarded encephalisation, but attached importance to the limitations to adaptive diversity implied by early acquisition of function of the forelimbs, and argued that this would stop the evolution of marsupial seals and bats. This argument is greatly weakened by the substantial forelimb modification in *Notoryctes* (Kirsch, 1977b).

Although it is easy to point to deficiencies in Muller's conclusions, we believe that acceleration of development to accommodate extrauterine existence may indeed set limits to the marsupial radiation by enforcing synchronous development and early acquisition of function in the upper half of the body. A re-examination of this issue is also appropriate because much of the original discussion took place before the relationship between ontogeny and phylogenetic diversification was clarified in an elegant review by Gould (1977). Alteration of the timing of development is called heterochrony, and is dependent on the dissociability of growth, differentiation and maturation. These changes may be expressed in terms of positive or negative changes in three parameters: onset of development, termination of development, which will often be associated with acquisition of function, and development rate (Gould, 1977, 1982; Alberch, Gould, Oster & Wake, 1979).

Table 3.5. *Differences in eutherian mammals giving birth to altricial offspring and precocial offspring*

Trait	Altricial type	Precocial type
Nest construction	Common	Unusual
Adult body size	Small	Medium or large
Homeothermy at birth	Imperfect	Well developed
Lower jaw development at birth	Imperfect	Well developed
Gestation	Short	Long
Litter size and teat count	Large	Small
Infant mobility	Low	High
Relative brain size	Low	High
Growth of brain after birth	Considerable	Moderate
Diel cycle	Nocturnal	Many diurnal species

After Martin (1975).

These changes may take a variety of forms, but all lead to one of two morphological expressions in descendants versus ancestors:

(1) Paedomorphosis, or the appearance of ancestral juvenile structures at later stages in descendants.

(2) Peramorphosis, or the passage through the ancestral adult stage to new extensions and modifications (similar to the phylogenetic extensions described above).

Although it is trivially obvious to say that various heterochronic alterations may interact to increase or cancel out the ultimate extent of any morphological change, it is conceptually easier both to consider the modes of change separately and to treat ontogeny as a linear function (see Fig. 3.16). Paedomorphic descendants can be produced by a decreased rate of differentiation (neoteny), an earlier termination of development (progenesis) or if the onset of development occurs late but the other parameters remain fixed (post-displacement). The converse changes give rise to peramorphic descendants, and are termed acceleration, hypermorphosis and pre-displacement, respectively.

The evolutionary importance of heterochrony has been underestimated, both because of the confusion surrounding the collapse of Haeckel's biogenetic law (ontogeny recapitulates phylogeny) (Gould, 1977), and because development was not considered explicitly in the modern evolutionary synthesis (Mayr & Provine, 1980; Gould, 1982; Stearns, 1982; see, however, Waddington, 1975). Gould (1977) regards peramorphosis as a motor of evolutionary process through addition to and refinement of organs, and forcefully argues that paedomorphosis represents an opportunity for escape from specialisation. While its origin within a lineage will be related to local environmental circumstances, the occurrence of paedomorphosis should increase the scope for future diversity. Further-

Fig. 3.16. The different modes of heterochrony, resulting from changes in (a) the rate of development, (b) the offset of development and (c) the onset of development. Circles denote the ancestral shape, filled stars denote the paedomorphic descendant shape, open stars denote the peramorphic descendant shape. After Alberch *et al.* (1979).

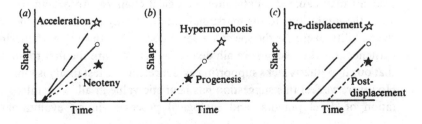

more, because heterochrony may result from a simple change to existing developmental processes, major evolutionary change can occur as a result of minor genetic changes, especially where the developmental trajectory is altered early in ontogeny (see e.g. Alberch, 1980, 1982; Arthur, 1982). Developmental constraints will also result in discontinuities in the range of phenotypes which can be expressed (Alberch, 1980, 1982).

Data on marsupial ontogeny are inadequate to place them clearly within this perspective. However, it does appear that Muller and Lillegraven overestimate the importance of developmental offset at the time of birth. In an elegant study of brain ontogeny in *Macropus eugenii* by Renfree *et al.* (1982), developmental trajectories resembled the eutherian pattern, with the rate of ontogeny changing abruptly at the time of acquisition of endothermy and thyroid function. This echoes the suggestion of Russell (1982a) that birth in eutherians is functionally similar to acquisition of endothermy in marsupials. There is also preliminary evidence suggesting that marsupial growth curves for closely-related species sometimes differ only in the timing of the offset signal. For example, the growth trajectories of *Petaurus breviceps* and *P. norfolcensis* are identical until about 100 days, when growth levels off in *P. breviceps* but continues unabated for a further 35 days in *P. norfolcensis* (M. J. Smith, 1979). Development in *Potorous tridactylus* and *P. longipes* is similar for about 13–14 weeks, but the hindfoot and tail continue to grow beyond this period in *P. longipes* (Seebeck, 1982). These results are quite different from the pattern in *Trichosurus vulpecula* and *T. caninus*, where genuine differences in growth rate occur (How, 1976).

Thus, the limitations to change through peramorphosis are less clearly established than was implied by Lillegraven & Muller. By contrast, the role of paedomorphosis may have been underestimated. One potential consequence of extrauterine existence may be the need to synchronise and accelerate developmental processes during early organogenesis, at least in the upper half of the body and in the alimentary tract. The opportunity for paedomorphosis through neoteny or post-displacement may therefore be greatly reduced, especially early in ontogeny when small ontogenetic changes may cause dramatic morphological changes. According to this view, the limitations to the marsupial radiation may be a consequence of the inability to escape the specialisation associated with early extrauterine existence, for the reasons we alluded to earlier. We are the first to admit that our hypothesis lacks supporting evidence, and its testability is limited. However, we feel this suggestion has heuristic value in allowing interpretation of developmental and ecological processes in an evolutionary framework. The unwillingness to attempt such syntheses may have harmed our understanding of both developmental and life history processes (Gould, 1982; Stearns, 1982). It may be, for example, that the occurrence of marsupial traits such as the migration of the urogenital ducts and precise pairing of arteriovenous loops in the brain (Johnson *et al.*, 1982), are associated with synchronised early ontogeny, a hypothesis which seems susceptible to further investigation. To engage momentarily in whimsy, it is even tempting to suggest that the potential for heterochrony should differ in the anterior and posterior parts of a marsupial. As the bipedality of macropods no longer seems explicable in terms of energy conservation (Thompson, MacMillen, Burke & Taylor, 1980), we lack an adequate hypothesis for macropod big-footedness!

4

Life histories of the carnivorous marsupials

It is now appropriate to examine the variation or adaptations which species exhibit within the constraints of being marsupial. Because of the effect of diet on growth and fecundity we will separate the marsupials into those species which feed predominantly upon animal tissue, the polyprotodont marsupials (this chapter), and those species which feed predominantly upon plant material, the diprotodont marsupials (Chapter 5). We have argued that the protein- and energy-rich diet of carnivorous species facilitates growth and reproduction, so in these taxa we might expect comparatively large litters and short periods of maternal investment, subject to the allometric constraints we have discussed previously. The abundance of some animal prey varies seasonally (e.g. arboreal insects; Chapter 2) and unpredictably (e.g. desert foliage-feeding insects) and we would expect this variability to have a strong influence in shaping the life histories of carnivores.

Dasyuridae

Despite their conservative body form and diet, the Dasyuridae have been exceptionally successful in Australia and New Guinea in the range of habitats occupied (Chapter 2), and this success is reflected in the wide range of life history strategies they exhibit. This diversity of life history strategies has aroused considerable interest because of the opportunities it offers for empirical analysis of life history theory (see Chapter 6). A review of these life histories by Lee, Woolley & Braithwaite (1982) recognised six distinct strategies among the thirty species for which there was information on patterns of reproduction and longevity. The strategies were distinguished on the basis of five characters (Fig. 4.1): the frequency of oestrus, the duration and the timing of male reproductive effort, the seasonality of breeding and the age at maturity. The information used for assigning some species in this classification was meagre, and the classification was consequently provisional. The following revision incorporates new information and considers 34 of the 52 dasyurid species. The order in which these strategies is discussed is not intended to imply an evolutionary sequence, but places species along a spectrum of life history strategies ranging from semelparity to iteroparity.

Life history strategies
Strategy I.

The first strategy (Strategy I) identified by Lee *et al.* (1982) was characterised by monoestrous females, an abrupt mortality of all males at the conclusion of their first short mating period, and a mating period which is shorter than gestation, highly synchronised and predictably timed. Nine, possibly ten species show this strategy (Table 4.1), and they are all small dasyurids (< 200 g). Although both *Antechinus swainsonii* and *A. stuartii* exhibit male die-off and synchrony in life history events, other life history traits (particularly female longevity) differ. At the time of writing we lack a complete analysis of either species, but data from each are often complementary, and so we use both species to illustrate this strategy.

Behavioural oestrus in captive *A. stuartii* lasts for an average of 6.2 days and in one instance lasted 13 days (Marlow, 1961; Woolley, 1966b). The precise duration of the period during which females will copulate in natural populations is unknown but may be of the order of two weeks, that is, roughly half of the length of gestation. Females become increasingly attractive to males as this period progresses and spontaneously ovulate at the end of behavioural oestrus, within a few days of one another (Woolley,

Fig. 4.1. Key factors in the classification of dasyurid life histories. After Lee *et al.* (1982).

Oestrous pattern	No. of seasons per ♂	Duration of mating period	Seasonality of breeding	Age at sexual maturity (months)	Strategy	No. of species
Monoestrous —— Annual —— Restricted —— Seasonal ——				11 ——	I ——	9 or 10
Perennial— Restricted—— Seasonal ——				11 ——	II ——	5
Monoestrous or polyoestrous —— Perennial— Restricted—— Seasonal ——				11 ——	III ——	4 or 5
Polyoestrous —— Annual—— Extended —— Seasonal ——				6 ——	IV ——	3
				8–11 ——	V ——	7
		Aseasonal ——		? ——	VI ——	3

Table 4.1. Life history traits of dasyurid marsupials

Strategy	Species	Adult body weight (g)		Month(s) of births	Usual litter size	Source
		♂♂	♀♀			
I	*Antechinus bellus*	55	34	Sept.		11, 58
	A. flavipes	56	34	July–Sept.	8–12	21, 53
	A. godmani	95	55	Aug.–Sept.		48, 59
	A. leo	70	36			47
	A. minimus	65	42	July–Aug.		49, 51
	A. stuartii	35	20	Sept.–Oct.	6–10	38, 52, 53
	A. swainsonii	65	41	July–Aug.	6–8	38, 56
	Phascogale calura	60	43	Aug.	6–8	7
	P. tapoatafa	199	145	Aug.	7–8	12, 18, 45
Inserta sedis	*Dasykaluta rosamondae*	25–40	20–30	Oct.–Nov.	6–8	59
II	*Parantechinus apicalis*	60–100	40–75	Apr.	8	54
	P. bilarni	12–44	12–34	Aug.–Sept.	5	8, 11, 59
	Pseudantechinus macdonnellensis	25–45	20–40	Aug.–Sept.	6	56, 59
	Satanellus hallucatus	400–900	300–500	July–Aug.	6–8	9, 23, 45
	Dasyurus geoffroii	1300	850	May–June	6–8	1, 4, 5
III	*Dasycercus cristicauda*	75–130	60–95	June–July	5–7	22, 37, 40, 55
	Dasyurus viverrinus	1300	880	May–Aug.	5–8	25, 29, 30, 33, 34
	D. maculatus	<7000	<4000	July–Aug.	5	20, 44
	Sminthopsis leucopus	15–25	15–25	Aug.–Oct.	10	43, 60
Inserta sedis	*Sarcophilus harrisii*	8000	6000	Mar.–May	4	10, 19, 30, 31, 35
IV	*Sminthopsis crassicaudata*	16–22	12–21	June–Feb.	6–8	28, 42, 56
	S. macroura	14–21	15–21	July–Feb.	6	27, 59
	S. murina	16–28	16–22	Aug.–Jan.	4–10	26

	Species			Breeding season		Sources
V	Ningaui sp.					
	Planigale m. maculata	6-10	6-13	Sept.-Jan.	6-7	17
	P. ingrami	4	7-9	July-Jan.	5-10	2, 3, 24, 41, 46
				Dec.-Mar. (NT) Sept.-sum. (Qd)	4-12	2, 32, 57
	P. tenuirostris	6	5	Summer	4-8	14, 15
	P. gilesi	11	7	July-sum.		14, 50
	Antechinomys laniger	30	20	July-Nov.	4	56, 59
	Dasyuroides byrnei	120	100	May-Jan.	5-6	6, 36, 39, 56, 57
VI	Antechinus melanurus	31-39	24-32	All months	3-4	16
	A. naso	44-63	37-48	All months?	4	16
	Planigale m. sinualis	8-22	7-15	All months?	4-12	2, 6, 13, 45

Sources: [1]Archer (1974); [2]Archer (1976); [3]Archer in Denny (1982); [4]Arnold (1976); [5]Arnold & Shield (1970a, b); [6]Aslin (1974); [7]A. J. Bradley, (personal communication); [8]Begg (1981a); [9]Begg (1981b); [10]Buchmann & Guiler (1977); [11]Calaby & Taylor (1981); [12]Cuttle (1982a); [13]Davies (1960); [14]Denny (1982); [15]Denny, Gibson & Read (1979); [16]Dwyer (1977); [17]Fanning (1982); [18]Fleay (1934); [19]Fleay (1935); [20]Fleay (1940); [21]Fleay (1949); [22]Fleay (1960); [23]Fleay (1962); [24]Fleay (1965); [25]Fletcher (1977); [26]Fox & Whitford (1982); [27]Godfrey (1969); [28]Godfrey & Crowcroft (1971); [29]Godsell (1982); [30]Green (1967); [31]Guiler (1970b); [32]Heinsohn (1970); [33]Hill & Hill (1955); [34]Hill & O'Donoghue (1913); [35]Hughes (1982); [36]Hutson (1976); [37]Jones (1923); [38]A. K. Lee & R. W. Martin (unpublished observations); [39]Mack (1961); [40]Michener (1969); [41]Morrison (1975); [42]Morton (1978a); [43]Read, Fox & Whitford (1983); [44]Settle (1978); [45]Taylor, Calaby & Redhead (1982); [46]Van Dyck (1979); [47]Van Dyck (1980); [48]Van Dyck (1982b); [49]Wainer (1976); [50]Whitford et al. (1982); [51]B. Wilson, (personal communication); [52]Wood (1970); [53]Woolley (1966a); [54]Woolley (1971a); [55]Woolley (1971b); [56]Woolley (1973); [57]Woolley (1974); [58]Woolley (1981); [59]P. A. Woolley (personal communication); [60]Woolley & Ahern (1983).

1966b; L. Selwood & M. P. Scott, personal communication). Females probably copulate with a number of males (see below), and since females are capable of storing sperm (Woolley, 1966b, Selwood, 1980), litters are possibly sired by more than one male. Abrupt mortality of all adult males (male die-off) begins at about the time of ovulation, and is complete within 5–10 days (Wood, 1970; McDonald, Lee, Bradley & Than, 1981). These events are synchronous from year to year. For example, in three years of a study of a population of *A. swainsonii* in the Otway Ranges, Victoria, male die-off occurred between 20 July and 30 July (Fig. 4.2). The causes of this mortality are discussed in Chapter 6.

The gestation period is approximately 28 days (Selwood, 1980), which is long for a marsupial of this size, and results from periods of stasis and slow early development (Selwood, 1980). Births occur 2–3 weeks after male die-off and are usually confined to a period of a few days in a particular population (Wood, 1970). Neonates attach to a teat following birth and remain attached for 30–50 days. The young are then suckled in a nest for a further 2–3 months and are weaned between December and February (Marlow, 1961; Wood 1970).

Fig. 4.2. The number of *Antechinus swainsonii* known to be alive on a trapping grid in the Otway Ranges, Victoria, showing the abrupt male die-off in July following mating. Cohorts born in August 1976 (stippled), August 1977 (cross-hatched), August 1978 (blank) and August 1979 (square-hatched). Data from A. K. Lee & R. W. Martin (unpublished observations).

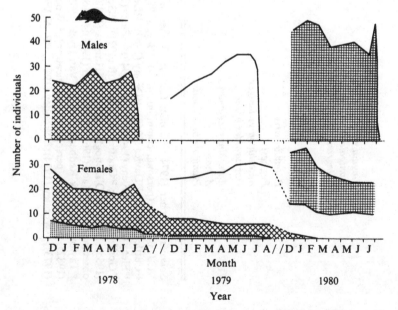

It was previously thought that litters of *A. stuartii* continued to use the maternal nest until autumn (April–June), when there was a reorganisation of the population (Braithwaite, 1979). However, by marking young attached to the teat, Cockburn, Scott & Scotts (1984b) have shown that the males of several populations of both *A. stuartii* and *A. swainsonii* in Victoria disperse at weaning. Females are more faithful (philopatric) to the maternal site, and share a nest with their mother and one or several unrelated males (Dickman, 1982a, Cockburn *et al.*, 1984b). On the basis of the overt aggression shown by males towards other individuals in contrived encounters at a neutral arena, Braithwaite (1979) concluded that individuals at his south-eastern Queensland study site probably nested alone after autumn. The set of males in this area could then be arranged in a dominance hierarchy. Subsequent trapping revealed that the dominant males in each year's cohort occupied roughly the same area of the study grid in two successive years. Braithwaite (1979) suggested that the moist soil and abundance of logs in this area fostered a greater abundance of arthropods than elsewhere, and that this contributed to the growth of the dominant males over winter; subordinates either showed no gain or lost weight. Consequently, the dominant males were heavier than subordinates when they entered the mating period and, he assumed, mated first and died first. Braithwaite (1979) concluded that females copulated only once or on a few occasions (see below) and that dominant males therefore did most of the insemination. There is now some evidence to the contrary.

Scotts (1983) used radiotelemetry to locate several nests of *A. stuartii* in a population in Victoria, and found that communal nesting by groups comprising both males and females continued until the time of male die-off. He found no evidence of male territoriality.

A. stuartii appears to be highly promiscuous and males may force copulation with a number of females (M. P. Scott, personal communication). Consistent with this observation is the conspicuous sexual dimorphism (Table 4.1) and the overt attempts by males to copulate with individuals of either sex at the time of mating. Observations of mating in the laboratory show that copulations consist of short bouts of thrusting interrupted by long periods of rest. Copulations usually last between 5 and 6 hours and may last up to 12 hours. Males and females housed together in captivity usually copulated once daily and up to 14 times over 13 days (Marlow, 1961; Woolley, 1966b). Females that are mounted frequently lose hair from the flanks and neck. Braithwaite (1979) did not observe this in female *A. stuartii* in the Mt Glorious population and concluded that females avoided repeated matings. However, in a population of *A. swainsonii* in Victoria we have subsequently found fur missing from the flanks and neck, and bite marks on the neck and rump of females (A. K. Lee & R. W. Martin, unpublished observations), and P. Woolley (personal communication) has found similarly-debilitated females in some populations of *A. stuartii*. These data suggest that a study of geographic variation in social organisation of *Antechinus* may be rewarding.

A notable feature of these life histories is the precision and synchrony with which events such as mating, male die-off, birth and dispersal occur within populations. We have already commented upon the synchrony of male die-off in a population of *A. swainsonii* in the Otway Ranges (Fig. 4.2). In a nearby population of *A. stuartii*, all females gave birth within 72 hours (A. Cockburn & M. P. Scott, unpublished observations). However, the timing of these events differs between populations and species (Table 4.1). Mating in *A. stuartii* occurs between late July and late September and in *A. swainsonii* between early May and late September (Fig. 4.3). Mating generally occurs later at higher altitudes in both species, later at low latitudes in *A. stuartii* and approximately a month earlier in *A. swainsonii* where it is sympatric with *A. stuartii* (Dickman, 1982b). Lee *et al.* (1977) suggested that the timing of mating in these species is set to synchronise the later stages of lactation and weaning with the flush of insects in late spring and summer. Late lactation was seen as the most

Fig. 4.3. Distribution of estimated time of mating of (*a*) *A. stuartii* and (*b*) *A. swainsonii*. Solid lines represent values where these species are sympatric and broken lines represent values where they are allopatric. After Dickman (1982b).

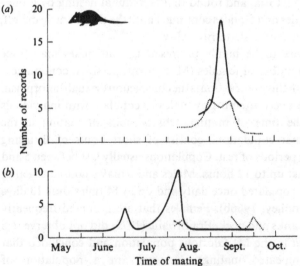

energetically demanding time for suckling females as they are then caring for litters with a combined weight that is several times their own. This is reflected in measures of field metabolic rates obtained by using doubly-labelled water (Table 3.1). Van Dyck (1982a) extended this hypothesis, arguing that the dominant influence on insect abundance was rainfall in the tropics, temperature in temperate areas and a combination of these variables in the subtropics. He was able to relate differences in the timing of reproduction in populations of *A. stuartii* and *A. flavipes* from different latitudes to conditions of temperature and rainfall. Differences in the timing of sympatric species have been attributed to different preferences for dietary items which peak at different times (Wainer, 1976) and to reproductive isolation or competition (Dickman, 1982b). Wainer (1976) based his hypothesis on the observation that *A. swainsonii* and *A. minimus* forage exclusively among soil litter and so take more larvae than the scansorial *A. stuartii* and *A. flavipes*. Seasonal peaks occur earlier among larvae than among adult arthropods (see Chapter 2) and thus the soil litter foragers, which were presumed to feed mostly upon larvae, should breed earlier than the scansorial species which were presumed to feed upon adult insects. Observations on the diet of sympatric species provide only weak support for this hypothesis (Hall, 1980b). Dickman's (1982b) argument was based upon purported increased differences in breeding times among sympatric versus allopatric populations of *A. swainsonii* and *A. stuartii*. Unfortunately, Dickman's (1982b) analysis was confounded because *A. swainsonii* is generally allopatric only in Tasmania (high latitudes) and at high altitudes, and *A. stuartii* is allopatric at drier sites and at low latitudes. Dickman's analysis was unsuitable for resolution of the relationship of sympatry to these climatic variables, and his hypothesis needs corroborative evidence. Neither of the available hypotheses adequately tackles the great variation in the timing of reproduction exhibited by *A. swainsonii* in comparison to *A. stuartii* (Fig. 4.3).

Since only a single litter is produced annually, and each young is permanently attached to one teat, variation in fecundity can only be achieved by adjusting either the number of neonates which attach to teats or the number of teats. Dasyurids produce a surplus of young at birth (Hill & O'Donoghue, 1913; Flynn, 1923; Woolley, 1966b; Hughes, 1982; Selwood, 1983), and there is a strong tendency for neonates to attach to all or all but one teat in species exhibiting Strategy I (Fig. 4.4). Variation in litter size is achieved largely by variation in teat number and occurs both within and between species (see Chapter 6). Litters are generally large (6–13 young/litter). The probability of a female surviving to produce two litters is relatively high in *A. swainsonii* (Cockburn, Lee & Martin, 1983), but in

A. stuartii most females produce only a single litter in a lifetime and are therefore semelparous (see Chapter 6). All females tend to produce a litter in each breeding season.

Strategy I is known from two genera, *Antechinus* and *Phascogale* (see Plate 4.1) which are confined to the forests, woodlands and heaths of northern, eastern and southern Australia, where the climate is both seasonal and predictable (Fig. 4.5). The strong association with predictable climates is not surprising as the strategy offers no insurance against reproductive failure other than breeding in a second year.

Strategies II and III.

The second and third strategies identified by Lee *et al.* (1982) grouped dasyurids which are monoestrous, or, if polyoestrous, tend to reproduce synchronously and once-yearly and so behave as if monoestrous (facultative monoestrous) (Table 4.1). In these species some individuals of both sexes survive to reproduce in at least two years. They include both small and large dasyurids (0.012– > 7 kg).

In the first of these strategies (Strategy II) females appear to be monoestrous. An example is provided by *Parantechinus bilarni*, studied by Begg (1981a) at Little Nourlangie Rock, Northern Territory. Here mating probably occurs in late June and early July, as births occur in July and August. At the time of mating, males range widely, leading to a high

Fig. 4.4. The frequency of different-sized litters in populations of *Antechinus stuartii* differing in teat number. (*a*) Otway Ranges, 6 teats; (*b*) Sherbrooke Forest, 8 teats; (*c*) Leonards Hill, 10 teats. Data from Cockburn *et al.* (1983).

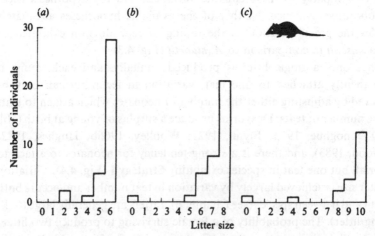

number of transients and low survival in the trapped population. In Begg's (1981a) study pouch young were first observed on 11 August, and were last observed attached to the teats on 18 September. Suckling continued until December and January, and free-living young were first trapped in December. There was a decline in the number of males trapped in February, which may reflect dispersal of male young shortly after weaning. Survival was also low during the mating period. Female survival is generally high and cohorts of females suffered only gradual attrition. The longest-surviving male *P. bilarni* were at least 32 months of age and females were 35 months of age. Males and females surviving to their second breeding season reproduced or appeared to be capable of reproduction.

Lee *et al.* (1982) placed three species in Strategy III, including *Dasyurus viverrinus*. Fletcher (1977) found that the female *D. viverrinus* were usually monoestrous, but some females underwent a second oestrus if they were unmated or prematurely lost the first litter. The importance of a second or later oestrus may vary between populations. In a Tasmanian population of *D. viverrinus* studied by Godsell (1982), mating mostly occurred within a confined period from mid-May to early June, but an analysis of birth dates of specimens collected by Hill and associates in New South Wales

Plate 4.1. *Phascogale tapoatafa*, the largest semelparous dasyurid. (Peter Cuttle.)

Fig. 4.5. The distribution of dasyurid species employing the life history strategies defined in Fig. 4.1. (*a*) Strategy I; (*b*) Strategy II; (*c*) Strategy III; (*d*) Strategy IV; (*e*) Strategy V. Numerals denote the number of species found within that portion of a 2° grid.

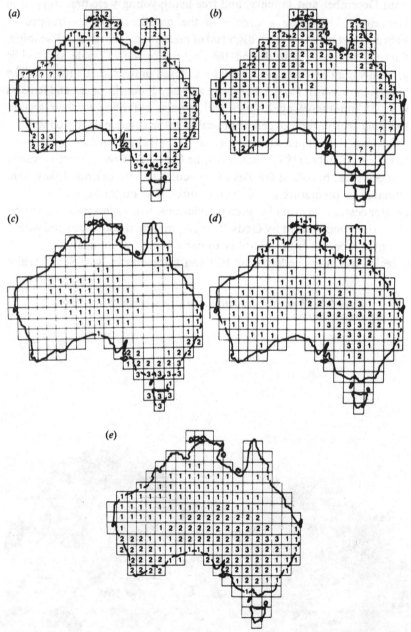

suggests a bimodal distribution with peaks in late June and late July (Tyndale-Biscoe & Renfree, in press), implying a mating period of a month or more.

In Godsell's (1982) Tasmanian population, litters of six young were carried in the pouch until mid-August and then left in a den. Weaning was complete by the end of October. A substantial proportion of adult females disappeared from this study area in summer at a time when juveniles were first entering the population. The number of transient females increased during this period and Godsell (1982) suggested that as females tend to be philopatric at other times and share home ranges with their litters, they may disperse in the face of competition from young. Following this period female survival was high until the next cohort was weaned. Males, on the other hand, showed a gradual attrition until after the mating period, when there was a substantial decline in their numbers. The incidence of male transients was highest during the mating period. Godsell's (1982) data, though incomplete for the life of any cohort, suggest that few adults survive beyond a second breeding season, which is again surprising in view of their size (0.7–1.8 kg).

The range of teat numbers (4–8) found in species with these strategies encompasses the lower two-thirds of the range for Strategy I species (Table 4.1). Once again there is a tendency for most teats to be occupied, although the frequency of litters where all teats are occupied may be lower than in Strategy I species. In two breeding seasons, 46.7 and 26.1% of female *P. bilarni* had the full complement of young shortly after birth (Begg, 1981a). In comparison, 55.5% of female *A. swainsonii* on a trapping grid in the Otway Ranges had the full complement of young in three consecutive seasons (Cockburn *et al.*, 1983).

There is no obvious set of attributes which are common to the habitats of species exhibiting these strategies, although most occur where the climate is less predictable than the areas inhabited by Strategy I species. *P. bilarni*, *Satanellus hallucatus* and *Pseudantechinus macdonnellensis* inhabit rocky areas in Arnhem Land or central Australia (Parker, 1973). *P. macdonnellensis* is also found associated with the giant mounds of the termite *Nasutitermes triodiae*, and *S. hallucatus* also occurs in eucalypt woodland. *Parantechinus apicalis* has been captured in heathland in south-western Australia (Woolley, 1977). Two Strategy III species, *D. viverrinus* and *D. maculatus*, are found in the forests and woodlands of south-eastern Australia that are inhabited by Strategy I species, whereas *Dasycercus cristicauda* inhabits sandhills and country surrounding clay-pans in central Australia (Parker, 1973).

In many respects the life histories of these species resemble those of Strategy I species. The principal difference is that some males as well as females breed in their second year (both Strategies II and III), and in those species which are facultatively monoestrous, the females are polyoestrous and so can replace a litter that is lost prematurely (Strategy III).

The position of the Tasmanian Devil (*Sarcophilus harrisii*) in this classification remains unclear. Matings appear to be synchronised and occur in March (Guiler, 1970b). The majority of births occur in April, and young are weaned in November and December (Hughes, 1982), indicating an exceedingly long period of lactation (7–8 months compared with 3–4 months in other dasyurids). Individuals survive to at least six years of age in the field (Buchmann & Guiler, 1977), which is again longer than the 1–3 years typical of other dasyurids for which longevity is known.

There is evidence of out-of-phase breeding (Green, 1967; Guiler, 1970b; Hughes, 1982), which suggests that the Tasmanian Devil is polyoestrous, although producing only one litter a year. The age at maturity may be two years in females (Fleay, 1952; Guiler, 1970b; Hughes, 1982) and not one year or less as in other dasyurids. There are four teats and litter sizes are accordingly small (Table 4.1).

Strategies IV–VI.

Lee *et al.* (1982) identified three further strategies among the remaining dasyurids, all of which are polyoestrous and usually produce at least two litters in a breeding season. Two of these strategies were characterised by an extended, seasonal breeding period (Strategies IV and V), and the third by year-round reproduction (Strategy VI). Strategies IV and V were distinguished on the time required to attain sexual maturity, approximately 6 months in the former and 8–11 months in the latter. The more rapid attainment of sexual maturity by species employing Strategy IV was seen to provide an opportunity for reproduction during the season of birth, and therefore a higher reproductive potential. However, there are reasons for suspecting that this advantage is not attained in natural populations. The time to maturity in Strategy IV species (6 months) is almost the duration of the breeding period, so that we would expect any attempt to gain advantage from reproduction in the season of birth to be accompanied by a burst of reproduction at the start of the breeding season. There is no evidence of this (Fig. 4.6). Further, Morton (1978a) found no evidence of reproduction in the season of birth in three natural populations of *Sminthopsis crassicaudata*, a species employing Strategy IV. Although it now seems unlikely that this difference is important in enabling some species to breed in the season of birth, it may reflect generally accelerated development in these species (Table 4.2), allowing adult females to produce and wean two litters in the period suitable for reproduction.

S. crassicaudata is the best-known polyoestrous and seasonally-breeding dasyurid and will be used to exemplify this strategy. In three populations of this species encompassing the transition from cool temperate grassland

Table 4.2. *Rates of development of pouch young of five small dasyurid marsupials, expressed as age in days at first appearance of feature*

Species	Ear buds	Ears open	Eyes open	Left alone in nest	Weaned	Source
Antechinus stuartii	20	65	62	40	90	Marlow (1961)
Sminthopsis murina	—	28	49	34	58–65	Fox & Whitford (1982)
Ningaui sp.	10	54	48	42–44	76–81	Fanning (1982)
Planigale gilesi	8	—	43	37	65	Whitford *et al.* (1982)
Dasyuroides byrnei	—	—	75	56	95–105	Aslin (1974)

Fig. 4.6. Distribution of birth dates of *Sminthopsis crassicaudata* at (*a*) Fowlers Gap, New South Wales, and (*b*) Werribee, Victoria. From Morton (1978a).

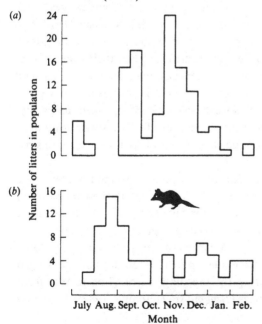

to desert shrubland, Morton (1978a) found that litters were produced over a six- to eight-month period between June and February (see Fig. 4.6). Females generally produced two litters in this period, one before and one after October.

Gestation requires only 12 days, less than half the length of gestation in *A. stuartii*, and weaning occurs at 60–65 days of age. Females can attain sexual maturity in about six months under laboratory conditions (Godfrey & Crowcroft, 1971). In natural populations, most individuals breed in only one breeding season and have a maximum longevity of about 18 months (Morton, 1978a). There is no abrupt post-mating mortality of males as occurs in *A. stuartii*.

The capacity of males for prolonged copulation may be less than in *A. stuartii*. In the laboratory, males may mount females for up to 11 hours with periods of rest between thrusting lasting no more than 7 minutes. Males may copulate for two or three successive nights, but, after the first night, for periods of approximately two hours only (Ewer, 1968). Contrived encounters between pairs of individuals showed none of the offensive aggression or changes in the levels of agonistic and amicable behaviour which occurred with the onset of breeding in *A. stuartii* (Morton, 1978b). Most nest groups found by Morton during the breeding season comprised a single male with either a proestrous or oestrous female. Ewer (1968) observed that males in captivity fought only in the presence of oestrous females. Individuals of both sexes inhabit overlapping home ranges which drift considerably over a period of time. This led Morton (1978b) to conclude that the social organisation was 'very loose' with no long-term bonds between individuals or between individuals and their range. These observations suggest that the mating system is mate-defence polygyny, and is not accompanied by the conspicuous sexual dimorphism in body size and frenetic mating behaviour so characteristic of the *Antechinus* spp. employing Strategy I. In view of the occurrence of only a single adult male with proestrous and oestrous females, it seems likely that litters are sired by only a single male.

Information now available for other species grouped under these strategies (IV and V) shows that there are differences in their tolerance of other individuals, in their rates of development (Table 4.2) and in their litter sizes (Table 4.1). The *Ningaui* sp. studied in captivity by Fanning (1982) showed intolerance between all conspecifics except for mating pairs, females with offspring and litter mates up to 120–150 days of age. Caged male *S. murina* also fight during the breeding season (Fox & Whitford, 1982). A number of these species (*Ningaui* sp. (Fanning, 1982); *S. murina*

(Fox & Whitford, 1982); *S. virginiae* (Taplin in Whitford, Fanning & White, 1982); *Planigale maculata* (Van Dyck, 1979); *P. gilesi* (Whitford *et al.*, 1982)) have mate-attracting calls which are usually emitted by both males and females when they are housed alone during the breeding season. In females, the intensity of calling increases with the onset of oestrus (Fox & Whitford, 1982; Read, Fox & Whitford, 1983). Males call throughout the breeding season, but most often during the first month (Read *et al.*, 1983). Fanning (1982) has suggested that mate-attracting calls provide an effective means of communication for solitary species living in habitats with isolated cover in which tactile communication may place animals at risk to predation.

Litter size in these polyoestrous species embraces the entire range for dasyurids. The largest litters and the greatest variation in litter size occur among the small planigales, and may be related to their propensity to use habitats such as flood plains (Denny, 1982), which are subject to frequent perturbation. The desert *Sminthopsis, Antechinomys laniger* and *Dasyuroides byrnei* have smaller litters which are less variable in size, suggesting that they may spread reproductive effort. There is considerable variability in teat number within populations of *S. crassicaudata* (6–11) (Morton, 1978a), but there is no evidence that this is reflected in different litter sizes which may offer advantages in a variable environment.

Three polyoestrous species, *Antechinus melanurus* and *A. naso* from New Guinea, and *Planigale maculata sinualis* from the north of the Northern Territory, appear to reproduce the whole year round (Dwyer, 1977; Taylor, Calaby & Redhead, 1982). We have no detailed information on their life histories.

Polyoestrous dasyurids with a prolonged mating period and a tendency to produce two or more litters during that period are found over almost the entire range of habitats utilised by the family. For example, *S. murina* occurs in open forest and in mid-seral stages of coastal heath regenerating after fire; the Strategy I species *A. stuartii* occupies the same sites during later seral stages (Fox, 1982c), *S. crassicaudata* is found in cool temperate grassland, desert shrubland and tussock grassland, and stony desert (Morton, 1978a; Morton, 1982), and the planigales occur in a variety of vegetation types in monsoonal New Guinea and north and eastern Australia, and in the semi-arid interior of Australia (Archer, 1976; Denny, 1982). Polyoestrous species predominate among dasyurids in arid Australia, where the family has been particularly successful (Morton, 1982). However, they are not found in the temperate closed forests of southern Australia where Strategy I species prevail.

Evolution of life history strategies. The distributions of species showing the different life history strategies provide clues to their evolution. We have seen that dasyurids which tend to produce one litter annually (Strategies I–III) are primarily confined to areas of Australia which have predictable and highly seasonal climates (Fig. 4.5). The accessibility of invertebrate food in these areas is least in winter and greatest in late spring and summer. The later stages of lactation and weaning in these species consistently occur in late spring and summer when the accessibility of food is most critical. Maternal commitment to young in mammals is usually greatest during late lactation (see Chapter 3) and readily accessible food is probably essential for weaning young that are naive in prey-capturing skills. In contrast to late lactation and weaning, there is considerable variation in the timing of mating, much of which can be accounted for by allometric effects on the period of maternal dependence (gestation and lactation). Larger dasyurids such as *D. viverrinus*, which have a long period of maternal dependence, mate earlier in the year than small monoestrous dasyurids such as *A. stuartii*.

Braithwaite & Lee (1979) suggest that the need to associate late lactation and weaning with the peak in food accessibility is probably the circumstance favouring the selection of synchronous oestrus and mating in these species. They go on to argue that an inability to raise two litters during the period of the year when food is readily accessible is the circumstance favouring the monoestrous condition and ultimately the evolution of Strategy I. They point out that if *A. stuartii* proceeds to raise a second litter after weaning the first in December or January, then lactation and weaning of the second litter coincides with autumn and winter when food is scarce. Significant failure of this second litter would favour selection of monoestrous females, particularly those which withhold their residual reproductive effort so that they can breed in the spring of their second year, or increased reproductive effort in their first breeding season. With these constraints on reproduction and the low chance of small mammals surviving to reproduce in a second year (see data on life expectancy in French, Stoddart & Bobek, 1975), selection should also favour males which maximise their reproductive effort during their first breeding season, even though they may reduce their longevity as a consequence. In Chapter 6 we present a hypothesis which implies that the abrupt post-mating mortality of males of Strategy I species is the consequence of increased corticosteroid activity during the mating period, facilitating increased reproductive effort.

Species employing Strategies II and III (males and females live to reproduce in a second year, but wean only one litter annually), are either larger than Strategy I species (mostly Strategy III species), and therefore have a greater life expectancy, or occur in habitats such as rock crevices and termite mounds which probably afford greater protection for foraging than the habitats occupied by Strategy I species. For these dasyurids, some of which occur where the climate is relatively unpredictable (e.g. *Pseudantechinus macdonnellensis*), the survival of some individuals of both sexes to reproduce in a second year provides two possible advantages: an opportunity to spread reproductive effort and so reduce the probability of loss of litters through overcommitment (they generally have small litters), and an insurance against failure at their first attempt at reproduction. Additional insurance against loss of a litter or failure to mate is found in species showing Strategy III, in which some females are polyoestrous, although only producing one litter annually.

The remaining strategies (Strategies IV–VI) are confined to small polyoestrous species which usually produce more than one litter annually. We have described three circumstances under which they predominate. Most of these dasyurids occur in the Australian arid zone (Fig. 4.5). Here resource abundance is profoundly influenced by rainfall, which is usually low and unpredictable but in some years is exceptionally high (see Newsome & Corbett, 1975; Morton, 1978a, 1982). Morton (1978a) argues that there may be temporary short supplies of food in this environment associated with drought and that these could cause loss of litters. He argues that repeated reproduction in a breeding season enhances the probability of reproductive success. Production of two small litters rather than a single large litter may also lower the probability of loss of litters. It is curious that, as a very successful group in the arid zone, these species do not reproduce opportunistically in response to rainfall, but retain the predominant dasyurid pattern of reproduction in spring and summer. Three species reproduce year-round (Strategy VI), and these occur in tropical habitats where there is probably an abundant supply of insects throughout the year.

Perhaps the most interesting occurrence of polyoestry is in *S. leucopus* and *S. murina* (Fox, 1982c). The dilemma here is that we would predict that the forests and heaths inhabited by these species would furnish a food supply which is predictable and highly seasonal, circumstances in which *A. stuartii* replaces *S. murina* in later seral stages in certain heathlands (Fox, 1982c), and also replaces *S. leucopus* as disturbed patches of forest regenerate (A. K. Lee & A. Cockburn, unpublished observations). While we have no information on the supply of food in the different seral stages which may help to resolve this dilemma, this example should remind us that the evolution of life history strategies is not governed by the seasonal distribution and predictability of the food supply alone. The high reproductive rate of these species may allow them to exploit patches of habitat which may only be suitable for as few as three years (Fox & McKay, 1981; Fox & Whitford, 1982), and to export young to colonise other patches of similar quality. Why they are able to utilise these patches and A. stuartii is not, remains to be resolved.

A surprisingly high percentage of the dasyurids we have discussed produce only a single litter annually. In fact exactly half of the 34 species classified here are monoestrous, and although this proportion will decline as information is obtained for the remaining 15 or so species, which are predominantly desert-dwelling or New Guinean species, and therefore likely to be polyoestrous, it is unusually high for a taxon of primarily small mammals (Asdell, 1964; Eisenberg, 1981). Braithwaite & Lee (1979) observed that the developmental times were long in marsupials when compared with similarly-sized eutherian mammals and that this would restrict the number of litters which could be produced in environments where there was only a short period suitable for reproduction. In Chapter 3 we have argued that this low fecundity is associated with the slow pace of growth in marsupials, which is particularly pronounced in small species. Even in those Sminthopsis spp. with relatively short gestation (12 days) and lactation (60–70 days), a minimum period of 5–6 months is required to produce two litters. While most small eutherians appear to spread reproduction over several litters during a breeding season, a single litter may be produced where the growing season is exceptionally short. Small mammals inhabiting mountain tops at middle latitudes appear potential candidates. For example, Vaughan (1969) found that the shrew Sorex vagrans and the rodents Thomomys talpoides and Eutamias minimus produced a single litter yearly at 3000 m in Colorado in comparison with two or more litters at low altitudes. Whether or not these species are semelparous will depend upon the probability of surviving to reproduce in a second year.

Despite our ability to identify the probable life history strategies of more than two-thirds of species in the Dasyuridae, our coverage remains superficial and largely speculative. Many of the characters used have been ascertained from laboratory colonies and we have no knowledge of their natural significance. Where information has been drawn from natural populations, it has often been drawn from a single population of each species; with the exception of three species, A. stuartii, A. swainsonii and

S. crassicaudata, we have no indication of the range of life history tactics available. Yet the diversity of strategies, and especially the occurrence of monoestrous species, synchronously- and predictably-timed life histories and semelparity, provides us with opportunities to investigate problems in evolutionary ecology not previously available in mammals. These are considered in Chapter 6.

Didelphidae

While we frequently lament the paucity of data that are available for Australian marsupials, life history and population processes in the New World species are even less well understood. This situation is so poor for caenolestids and *Dromiciops* that we are unable to discuss them. The species for which most information is available is *Didelphis virginiana*, which probably had a recent origin and range expansion (Gardiner, 1973). This recent range expansion and the enormous range now occupied limit our ability to generalise to other species or to analyse evolution in life history traits. Hunsaker (1977) reviews the biology of this species in some detail, and we do not propose to elaborate on his discussion, other than to place the biology of *D. virginiana* within the perspective of the remainder of the Didelphidae.

The chief problem with seeking generalities or pattern within the didelphids is the narrow geographical ambit of population studies. While comparative studies on genera other than *Didelphis* have been made in Panama (Enders, 1935; Fleming, 1972, 1973), Venezuela (O'Connell, 1979), French Guyana (Charles-Dominique *et al.*, 1981; Atramentowicz, 1982), and Brazil (Davis, 1945a, b; Miles, de Souza & Povoa, 1981; Streilein, 1982a, b), they all focus on tropical communities, thus preventing inclusion of the southern or high altitude species. This may cause underestimation of the diversity of life history patterns (particularly among the diverse genera *Marmosa* and *Monodelphis*), and preclude comparison with Australian species, where the best population data are from temperate forms.

Within these constraints, we may conclude that most didelphids studied to date are seasonal breeders, polyoestrous and polytocous (Table 4.3), though *Monodelphis domestica* may approach year-round reproduction in the Caatinga of Brazil (Streilein, 1982a). Litter size varies between habitats and with latitude (O'Connell, 1979) (Fig. 4.7), confounding attempts at comparison between sites. We discuss geographic variation in litter size in Chapter 6, using the dasyurid genus *Antechinus* to dissect current theories pertaining to such gradients. The best data describing litter size differences among sympatric didelphids are those of Charles-Dominique (1983) for a group of five species in French Guyana (Fig. 4.8). The small *Marmosa* species have larger litters than other genera. These data also provide some support for the contention of Eisenberg & Wilson (1981) that arboreal species have smaller litters than terrestrial/scansorial species when allometric factors are taken into account, but more data of this type are required to confirm this trend.

The number of litters produced each season varies between one and four, and most authors have speculated that the duration of the breeding season and the number of litters successfully weaned are controlled by the availability of food and the period between successive litters. The latter parameter is correlated with body size (Chapter 3), while in neotropical habitats seasonal variation in the availability of food is generally related to dramatic annual variation in rainfall as temperature regimes are equable. Charles-Dominique *et al.* (1981) and Atramentowicz (1982) provide the most convincing evidence of a direct connection between food and reproduction in a study of *Caluromys philander* and *Philander opossum* in French Guyana. Both species eat fruit, arthropods and nectar, and *C. philander* may supplement its diet with gums (see Chapter 2). The availability of fruit is restricted between June, at the end of the wet season, and November, at the beginning of the dry season. The availability of insects is also restricted during the early part of this period, and fluctuates more sharply in the understorey, where *P. opossum* forages, than in the

Fig. 4.7. Variation of litter size with latitude in *Didelphis* spp. Filled circles, *D. virginiana*; open circles, *D. marsupialis* north of the equator; filled squares, *D. marsupialis* south of the equator; triangle, *D. albiventris* north of the equator; open square, *D. albiventris* south of the equator. Data from O'Connell (1979), Streilein (1982a) and Charles-Dominique (1983).

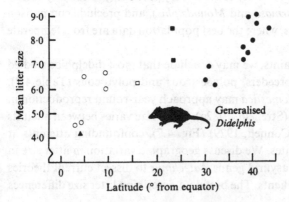

Table 4.3. *Life history traits in the Didelphidae. All species are seasonal breeders, polyoestrous and polytocous*

Species	Litter size	Litters per year	Interval between litters (days)	Age at maturity (months)	Longevity in wild (months) ♂♂	Longevity in wild (months) ♀♀
Caluromys derbianus	3.0[1]	—	—	—	—	—
C. philander	4.1[2]	1–2[2]	150[2]	9.5[2]	—	—
Chironectes minimus	3.5[1]	1–3[3,4]	—	—	—	—
Didelphis albiventris	See Fig. 4.7	—	—	—	—	—
D. marsupialis	(See Fig. 4.7)	2–3[1]	110[1]	6.2[1]	—	—
D. virginiana	(See Fig. 4.7)	2–3[1]	110[1]	—	24[1]	24[1]
Marmosa cinerea	6.3[2]	—	—	—	18[5]	9.3?[5]
M. fuscata	6.0[5]	1–2[5]	—	—	—	—
M. murina	8.5[2]	—	—	—	—	—
M. robinsoni	10[6], 14[5]	1–2[5]	85[1]	9.0?[1]	8.1[5]	9.5[5]
Metachirus nudicaudatus	5.0[7]	—	—	—	—	—
Monodelphis brevicaudata	7.0[5]	—	—	—	10.5[5]	13[5]
M. domestica	7.3[8]	—	—	6.0[1]	—	—
Philander opossum	3.4[7], 4.2[2]	2–4[2]	90[2]	5.0[2]	—	—

Sources: [1]Hunsaker (1977); [2]Charles-Dominique *et al.* (1981); [3]Streilein (1982a); [4]Tyndale-Biscoe & MacKenzie (1976); [5]O'Connell (1979); [6]Fleming (1973); [7]Eisenberg & Wilson (1981); [8]Streilein (1982b).

canopy, where *C. philander* is most active. This food shortage is reflected in a decline in weight and fat reserves in both species, and a decline in reproduction. A rapid recovery of reproduction occurs among both these species when prolific fruiting occurs at the end of the dry season (Fig. 4.9). The precise onset of reproduction and knowledge of the time between successive litters enabled Charles-Dominique *et al.* (1981) to estimate both pouch survival and weaning success for successive cohorts of young after the upsurge in reproduction. *P. opossum* shows high pouch survival (> 95%) for the first two litters, but low success in the third and fourth litter (28%). This is reflected in recovery of young (> 17% for the first two litters, 0% subsequently). *C. philander* has some success with the first litter (98% survive pouch life; 17% are weaned), but the second litter usually fails (18% and 0% respectively). Reproductive failure is associated with starvation of mothers and their young, which leads both to prolongation of lactation and to cessation of reproduction (Atramentowicz, 1982). Streilein (1982a) suggests that restricted rainfall in semi-arid habitat in the Brazilian Caatinga may restrict *D. albiventris* to production of a single litter annually, though this species may be polyoestrous elsewhere (Tyndale-Biscoe & MacKenzie, 1976). The results for *P. opossum* may explain why other didelphids are often reported to produce two litters in a breeding season (Table 4.3), although support for this conclusion is generally less substantial than that described above.

A further problem raised by these data is the lower annual fecundity of *C. philander* in contrast to that of *P. opossum*. This lower fecundity is

Fig. 4.8. Litter size in sympatric didelphids at Cabassou, French Guyana. Filled circles, terrestrial/scansorial species; open circles, arboreal species. Data from Charles-Dominique (1983).

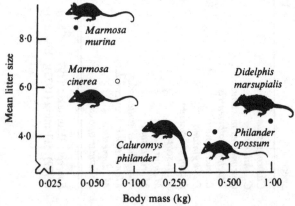

related to the greater time required to produce a litter (Table 4.3), and this is associated with the long duration of lactation in *Caluromys*, which needs to be considered in the light of results reported by Eisenberg & Wilson (1981). They suggest that *Caluromys* is distinguished from other didelphids by high encephalisation quotients, high basal metabolic rates, high longevity and small litters. These results require qualification. McNab (1978a) provides two measures of basal metabolic rate for *C. derbianus*, and only one of these is high for marsupials, so it would be premature to distinguish *Caluromys* on this trait. As discussed above, the difference in litter size between *Didelphis*, *Philander* and *Caluromys* is rather slight when macrogeographic factors are excluded. Any difference in fecundity appears to stem from the prolongation of lactation. We must consequently restrict discussion of life history traits which distinguish *Caluromys* to high encephalisation, long duration of lactation and high longevity. Although Eisenberg & Wilson (1981) used animals in captivity to assess relative longevity, the distinction of *Caluromys* on these grounds is supported by the field observations that *Caluromys* has a higher probability of breeding in more than one season than *Marmosa*, *Monodelphis* and *Philander* (O'Connell, 1979; Charles-Dominique *et al.*, 1981).

Fig. 4.9. Percentage of lactating female didelphids at Cabassou, French Guyana. (a) *Philander opossum*; (b) *Caluromys philander*. (c) Production of fruit pulp (kg dry weight) by the twelve dominant tree species in an area of 8.5 ha. After Charles-Dominique *et al.* (1981).

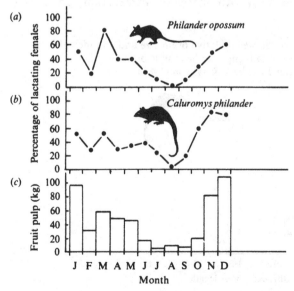

Eisenberg & Wilson (1981) have suggested that an arboreal existence promotes large brain size and a correspondingly large cerebellum (see Fig. 4.10). They further suggest that this increased encephalisation among mammals is part of an adaptive syndrome which includes increased longevity, decreased litter size and sometimes an increased percentage of the juvenile life span spent in social learning (see also Eisenberg, 1981). Eisenberg & Wilson (1981) concede several problems with their hypothesis, and suggest that preference for transitional habitat by the purportedly arboreal genus *Metachirus* fails to facilitate the greater encephalisation observed in the similarly arboreal *Caluromys*. However, occupancy of mature habitat by *Caluromys* leads to trophic competition with both closely-related and distantly-related taxa, with consequent additional selection favouring increased parental care, reduction of litter size and increased potential life span. They suggest that this combination of influences has caused *Caluromys* to converge toward the life history of nocturnal arboreal prosimians.

This argument is weakened by the absence of marked litter size differences between arboreal and terrestrial species, and by the conservative and simple social organisation of all Didelphidae in contrast to the prosimians (Miles *et al.*, 1981; Charles-Dominique, 1983). Further, Charles-Dominique *et al.* (1981) point out that all the frugivorous didelphids they examined were most abundant in early stages of forest succession in French Guyana, and Miles *et al.* (1981) suggest that *Metachirus* is almost exclusively terrestrial in Brazil. An alternative explanation of the long

Fig. 4.10. Relation between relative brain size and an index of arboreality (ratio of tail-length to head plus body-length) in Didelphidae. After Eisenberg & Wilson (1981).

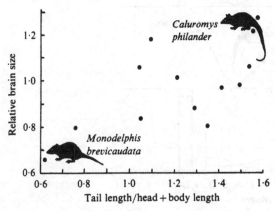

duration of lactation in *Caluromys* might be the diet of this species. Atramentowicz (1982) showed that this species relies more on plant material than do other didelphids, eating fruit and exudates such as nectar and gums. *Caluromys* has a well-developed caecum and colon relative to other didelphids examined; this probably reflects selection for facilitation of bacterial degradation of plant tissue (Hladik, 1979; Charles-Dominique *et al.*, 1981). In Chapter 3 we showed that large herbivorous species have slower development and longer periods of maternal investment than insectivores or carnivores. We consider that the restrictions associated with herbivory are more likely to be the cause of prolonged lactation in *Caluromys* than are interspecific competition and the consequences of increased encephalisation. The high reproductive failure in the second litter produced by *C. philander* supports the viewpoint that the lengthy period of maternal investment should be viewed as a constraint rather than as part of a general adaptive syndrome.

In summary, analysis of didelphid life histories is constrained by the limited range of habitats from which population data have been collected. All species living in tropical habitats are polyoestrous, polytocous and seasonal breeders. Decline in food supply causes reproductive failure and inhibition. Several authors have pointed out that *Caluromys* spp. have lower fecundity than other didelphids as a result of the prolongation of lactation. This low reproductive rate probably stems from a herbivorous diet rather than being part of a general adaptive syndrome associated with arboreality in stable environments as has been postulated elsewhere. Comparatively high encephalisation occurs in arboreal species, but probably reflects requirements of locomotion in three dimensions as social behaviour among didelphids appears to be simple and similar among both arboreal and terrestrial species (Charles-Dominique, 1983).

Peramelidae and Thylacomyidae

Despite the range of habitats occupied by peramelids and thylacomyids (bandicoots and bilbies), all species are broadly similar in litter size, food habits, body size, rate of growth and duration of parental investment (Table 4.4; Chapter 3). All are polyoestrous and produce several offspring in a litter, and several litters in a year. The causes of this conservative variation are uncertain, but bandicoots warrant discussion in this chapter because they take less time to produce and wean a litter than any other marsupial. This comparison is particularly pronounced when allometric features are taken into account. The brief period of maternal investment appears to result from three distinct features: an extremely brief

Table 4.4. *Life history traits in the Peramelidae. All species are polyoestrous and polytocous*

Species	Locality	Breeding season	Litter size	Litters per year	Recruitment[a] (%)	Age at maturity (months) ♂	♀	Home range (ha) ♂	♀	Source
Isoodon macrourus	Queensland	Year-round (declines in May)	3.38	—	13.3, 7.1	—	—	2.8	1.5	Gordon (1974)
	Queensland	June–Mar.	2.90[b]	—	16.5	—	—	—	—	Gemmell (1982)
I. obesulus	Tasmania	July–Feb.	2.80	—	—	4.5	3.0	—	—	Heinsohn (1966)
	Victoria	Aug.–Jan.	3.04	2.2	15.0	—	—	5.3	2.3	Stoddart & Braithwaite (1979)
Perameles gunnii	Tasmania	June–Jan. (drought) June–Apr. (good year)	2.33	3.8	—	—	—	25.0	2.2	Heinsohn (1966)
	Victoria	Year-round (stops in drought)	2.30	—	—	—	—	—	—	Seebeck (1979); P. D. Brown (personal communication)
P. nasuta	New South Wales	Year-round (declines in late summer)	—	—	—	—	—	—	—	Close (1977)
	New South Wales	Year-round (less frequent in autumn)	—	—	—	—	—	—	—	Lyne (1964)

[a] Recruitment of locally-born young into the population.
[b] See also Fig. 3.5.

gestation, weaning young at a small size and acceleration of growth (see Chapter 3). The existence of these features in association with other aspects of bandicoot reproduction which are unique among marsupials requires further investigation. These other features include chorioallantoic placentation coupled with highly invasive chorionic villi and prolongation of the life of the corpus luteum of pregnancy beyond parturition (Gemmell, 1979, 1981; Renfree, 1980; Rothchild, 1981). The combination of these features strongly suggests that ancestral bandicoots underwent selection for high fecundity and rapid maturity, in contrast to other marsupials. Unlike most other species, bandicoots frequently attain reproductive maturity in the season of their birth (Table 4.4).

Variation in reproductive output is largely governed by variation in the length of the breeding season (Table 4.4). While some populations have individuals which give birth in all months of the year (e.g. *Isoodon macrourus* (Gordon, 1974); *Perameles gunnii* (P. D. Brown, personal communication)), births are less frequent in some months of the year (often late summer or early autumn). The onset of breeding after anoestrus may be gradual or abrupt and highly predictable from year to year (Fig. 4.11). The roles of photoperiod and food in generating this pattern are poorly understood. By contrast, there are convincing data implicating drought and declining food availability as determinants of the cessation of breeding. Heinsohn (1966) showed that births in a Tasmanian population of *P. gunnii* stopped as early as January in a drought year, but persisted until May when conditions were more benign. P. D. Brown (personal communication) noted complete cessation of breeding in a mainland population of *P. gunnii* during a severe drought, with births occurring 17 days after widespread heavy rain terminated the drought. The decline in breeding in *P. nasuta*

Fig. 4.11. Breeding pattern of *Isoodon obesulus* at Cranbourne, Victoria. After Stoddart & Braithwaite (1979).

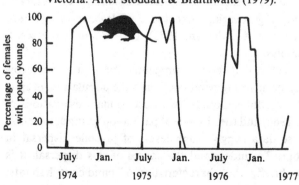

and *I. obesulus* populations in summer noted by Close (1977) and Stoddart & Braithwaite (1979) may also be associated with the effect of the drying of soil on prey numbers or accessibility.

Although there is only slight interpopulation variation in litter size, several aspects of intrapopulation variation warrant mention and further investigation. As we have mentioned previously, Gemmell (1982) demonstrated progressive litter reduction by captive *I. macrourus* as lactation progresses (Fig. 3.5). Although this effect was not excluded in other analyses, Heinsohn (1966) argued that the second and third litters produced by *P. gunnii* within a season are larger than the first and fourth litters. Stoddart & Braithwaite (1979) showed that litter size in *I. obesulus* is correlated with body weight, and that litters produced late in a breeding season are larger than those produced early in that season. Litter size in bandicoots may also be influenced by the rapid succession of litters. Heinsohn (1966) and Lyne (1974) present both empirical and photographic evidence that bandicoot neonates may be too small to attach to recently-used teats, and we might predict as a consequence that the number of young suckled in two litters produced in quick succession should not exceed eight, which is the number of teats found in all bandicoots (Lyne, 1964). A study which analysed the relative contribution of these factors to bandicoot litter size should be extremely rewarding.

Most bandicoots are solitary, and there is general agreement that local concentration of individuals in the wild reflects the dispersion of food resources rather than social bonding (see e.g. Heinsohn, 1966; Gordon, 1974). Even juveniles do not associate with their mothers for long periods (Gordon, 1974). Males range over larger areas than females (Table 4.4) and the mating system is probably promiscuous or polygynous. In the Tasmanian population of *P. gunnii* studied by Heinsohn (1966), the male ranges were ten times larger than those of females (Table 4.4), an observation which requires further investigation. Populations of *Perameles* exhibit greater home range overlap than those of *Isoodon* or *Macrotis* (Heinsohn, 1966; Johnson & Johnson, 1983), and *Macrotis* use scent-marking around burrows.

Several authors (Table 4.4) have commented upon the very low recruitment of locally-born young bandicoots into the population, in spite of the high fecundity of adults. Coupled with occasional observations of dispersal (Heinsohn, 1966) and the absence of parent–offspring association (Gordon, 1974), these data suggest high levels of juvenile dispersal in bandicoot populations. To understand the causes of this dispersal, it is necessary to review briefly the characteristics of bandicoot habitats.

Bandicoots appear to be capable of exploiting alienated areas such as farmland and rubbish tips (Heinsohn, 1966; Gordon, 1974; Seebeck, 1979), and even in areas of natural vegetation appear to prefer ecotonal (see e.g. Gordon & Lawrie, 1977) or early seral habitats (Stoddart & Braithwaite, 1979). This latter preference is associated with a high abundance of beetle larvae, a preferred prey type (Fig. 4.12). Bandicoots are apparently capable of rapid movement into recently burnt (e.g. Heinsohn, 1966) or cleared habitat (Stoddart & Braithwaite, 1979). In one instance a population decline was arrested and a rapid increase in numbers initiated by experimental clearing (Fig. 4.13). Stoddart & Braithwaite (1979) suggested that bandicoots are well adapted to use regenerating heathland because of their high reproductive rate and their utilisation of suboptimal adjacent habitats from which the optimum habitat can be reinvaded.

Cockburn, Braithwaite & Lee (1981) used data from populations of the heath rat, *Pseudomys shortridgei*, to examine the evolution of the life histories of small mammals in Australian heathlands in greater detail. Heathlands are developed on nutrient-poor soils, and productivity is always low, but highest early in succession, when nutrients bound up in plants are released into the soil. Because the suitability of any patch will

Fig. 4.12. The number of Melolonthine larvae (Scarabaeidae) in five soil cores (471 cm³ each) in 10- (cross-hatched) and 15-year-old (open) heath inhabited by *Isoodon obesulus*. Data from A. Opie (personal communication).

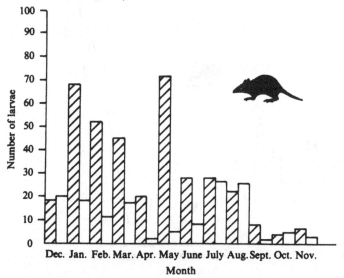

be ephemeral to those species (like *I. obesulus* and a number of *Pseudomys* spp.) that are dependent on a certain seral stage, the relative adaptiveness of breeding tactics will vary markedly through time. In a recently-colonised patch, the local environment will be at an earlier seral age than the mean age of other surrounding patches. As fire burns surrounding areas and the local patch matures, this situation will eventually be reversed. Additionally, once the local patch is saturated, the probability of producing large numbers of young (or any young) which can find an unoccupied home range within that patch is very low. Consequently, *Pseudomys* spp. exhibit two distinct phenotypes, which Cockburn *et al.* (1981) called annual and perennial types. Although the Peramelidae may be more long-lived, we suspect that the bandicoots described by Stoddart & Braithwaite (1979) behave in a similar fashion (A. M. Opie, personal communication).

The annual phenotype is common in animals living in recently-colonised areas, where habitat quality is high, as post-fire productivity and nutrient availability are at a maximum. On regenerating heathland, *P. shortridgei* adults disappear from their territories during or after the spring/summer

Fig. 4.13. Density of *Isoodon obesulus* (captures per 100 trap nights), at Cranbourne, Victoria, through a ten-year period. Data from Stoddart & Braithwaite (1979), A. M. Opie, B. Lobert, A. K. Lee & R. W. Martin (unpublished observations). The arrow indicates experimental clearing of part of the study area to stimulate regrowth.

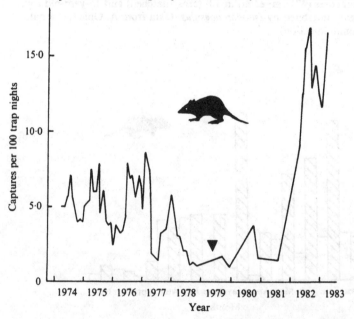

breeding season, and the local area is recolonised by young born during the season. The animals grow rapidly, occasionally breed in the year of their birth (Table 4.5), and are relatively short lived (Fig. 4.14). By contrast, in deteriorating habitats (e.g. mature heathland) where productivity is declining, adults hold sites tenaciously and live for three or four years (perennial phenotype) (Fig. 4.14; Table 4.5). Local juvenile survival

Table 4.5. *Temporal changes in life history traits of* Pseudomys shortridgei

	Annual type	Perennial type
Habitat characteristics	Early seral	Mature
Size of adult	Small	Large
Local juvenile survival	High	Low
Local adult survival	Low	High
Breeding in year of birth	Yes	No
Growth	Very rapid	Rapid
% surviving to breed	High	Low
Number of litters per season	One or two	Generally one

After Cockburn *et al.* (1981).

Fig. 4.14. Percentage of *Pseudomys shortridgei* surviving to various ages in (*a*) regenerating and (*b*) mature heathland. Solid line, males; dotted line, females. After Cockburn *et al.* (1981).

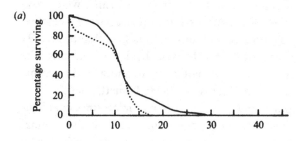

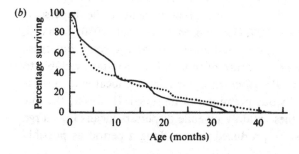

is very low, and generally depends upon the death of a resident adult. This corresponds with the three-year wave of replacement in male *I. obesulus* described by Stoddart & Braithwaite (1979), but the synchrony they observed may have been unusual, and the pattern did not continue in a later study at their site (A. M. Opie, personal communication). The disappearance of the young is associated with dispersal in *Pseudomys*, as young are caught away from their home range and show high levels of tail-scarring, a useful index of lost aggressive encounters (Cockburn *et al.*, 1981).

Thus it appears that adults switch from maximising local survival of their young to forcing their young to disperse and colonise new patches of microhabitat. As the availability of these areas is unpredictable, being dependent upon the vagaries of fire behaviour, it is advantageous to spread reproduction over as long a period as possible in order to minimise the probability of leaving no young at all. This is the inevitable consequence of enforcing dispersal in a year when no new patches of microhabitat become available. Such a temporally dynamic strategy is well adapted to repeated colonisation and exploitation of patches of newly-burnt heathland. It illustrates how the temporal and spatial variability of habitat may produce a counterintuitive result, that is, that local adult survival is highest in the poorest habitat. It is important to stress the term local survival as Cockburn *et al.* (1981) were unable to distinguish whether the disappearance of adults in regenerating habitat resulted from death or dispersal.

Evidence for dispersal comes from the microtine rodents, which also exhibit a temporally dynamic life history strategy (Cockburn & Lidicker, 1983). In crowded and overgrazed habitats, adults are philopatric. By contrast, in rapidly ameliorating or otherwise benign conditions, the mother abandons the brood nest to the young on weaning, and attempts to disperse to a new breeding site (Stenseth, 1978; Jannett, 1980).

These observations from *Pseudomys*, *Isoodon* and microtines enable us to examine one of the most perplexing questions facing small mammal ecologists. As dispersal appears to lower the probability of survival, why is it so frequent, and does it have some adaptive value to the dispersers? Recent reviews (Anderson, 1980; Gaines & McClenaghan, 1980; Tamarin, 1980) have been sceptical about advantages to the disperser, but these reviews ignore the ephemeral nature of much of the habitat available to small mammals. If, as in *Pseudomys* and bandicoots, local extinction is inevitable, there is an obvious premium on dispersal. However, the low survival probability of these mainly juvenile dispersers means that large numbers of young must be produced over as long a period as possible

(Giesel, 1976; Cockburn *et al.*, 1981). In this respect, the high fecundity of bandicoots made possible by their rapid development preadapts them to exploitation of the heathland environment. By contrast, other species of marsupials may be restricted in their ability to utilise heathland, and are therefore prone to frequent local extinctions (Braithwaite, 1979). In Australia it is certainly true that most of the species exclusively associated with heathland communities are rodents (Braithwaite, Cockburn & Lee, 1978; Kikkawa, Ingram & Dwyer, 1979; Newsome & Catling, 1979; P. R. Campbell & B. J. Fox, personal communications).

In summary, the Peramelidae exhibit little variability in reproductive traits in contrast to the Dasyuridae. Variation in life history tactics is principally facultative, and is manifested through changes in the length of the breeding season and the production of dispersive offspring. The brief period required to produce a litter may have preadapted bandicoots for habitats where high fecundity and dispersal are useful traits, enabling some species to occupy habitat not accessible to other marsupials.

Summary

Each of the three groups we have examined occupy many habitats across a broad geographic range. While the Peramelidae (bandicoots) are rather conservative in the range of reproductive traits they exhibit, the Dasyuridae show a variety of life history strategies that can be explained by current life history theory. The bandicoots mainly show facultative changes, while the life histories of the dasyurids are more rigidly controlled. The Didelphidae are too poorly known to enable us to place them within this perspective, but demographic studies of populations at high altitudes or high (southern) latitudes should resolve this uncertainty.

5

Life histories of the herbivorous marsupials

In this chapter we consider the life history strategies of diprotodont marsupials, marsupials which primarily depend upon plant tissues and plant exudates for food. In the previous chapter we assumed that the quality of food resources of marsupials which feed upon animal tissues was uniformly adequate and concluded that differences in the seasonality and predictability of supply were important in shaping their life history strategies. Here, however, we are concerned with animals whose food resources differ and vary markedly in quality as well as in supply (see Chapter 2), and for this reason we have chosen to review the life histories of diprotodont marsupials according to their feeding strategies.

Feeding strategies and life histories
Nectarivore

The two families (Tarsipedidae and Burramyidae) represented in this category appear to have similar life history traits (Table 5.1). These small pygmy possums and gliders are the most fecund of the diprotodonts, usually producing two or three litters a year, each of two to six young. They are unusual among marsupials in that some species breed in the season of birth.

Among these nectarivores the life history of *Tarsipes rostratus* is best documented. Births occur year-round in *T. rostratus* (Fig. 5.1), which is found in heathlands where flowering phenologies ensure substantial supplies of nectar and pollen in most months (Scarlett & Woolley, 1980; Wooller *et al.*, 1981). Some females produce at least two, and probably three litters a year. There is a nadir in females carrying pouch young in December, when flowering is least, and a peak in February–March, when flowering increases. Subsequent peaks in the percentage of females with

Table 5.1. *Life history traits of marsupial nectarivores (N) and frugivore/granivores (F/G)*

Species	♀ adult body weight (g)	Litter size	Litters/year	Annual fecundity	Month(s) of births	Age at weaning (months)	Age of ♀♀ at birth of first litter (months)	Breeding in season of birth	Natural longevity (years)	Source
Tarsipes rostratus (N)	8–19	2–3 (1–4)	2–3	4–9	All months	3	6	Yes	<1	Wooller et al. (1981)
Cercartetus concinnus (N)	13	4 (3–6)	—	—	—	—	12–15 (?)	—	—	Bowley (1938); Cassanova (1958); Clark (1967)
C. nanus (N)	24	4–5	2	8–10	Sept.–Apr.	—	5	Yes	>1.8	V. Turner (personal communication)
C. caudatus (N)	30	2 (1–4)	—	—	Aug.–May	2–3	—	—	—	Dwyer (1977); Atherton & Haffenden (1982)
Acrobates pygmaeus (N)	10.5–16.5	2–3 (1–4)	2	4–6	July–Jan.	—	6–8	Yes	>2	H. Frey & M. Fleming (personal communication)
Burramys parvus (F/G)	40	4	1	4	Nov.	1.5	12	No	>4	Dimpel & Calaby (1972); Mansergh & Walsh (1983); A. Kerle & M. Fleming (personal communication)

pouch young occur in May–July and September–October, at intervals of approximately 90 days, that is, the minimum interval between birth and weaning of young (Wooller *et al.*, 1981). Females may conceive postpartum, and the blastocysts resulting from conception at this time remain quiescent in the uteri while the previous litter is suckled (Renfree, 1981). Young leave the pouch at about 60 days of age, but reactivation of development of the blastocysts is slow and births occur approximately 90 days after conception. The ecological significance of embryonic diapause in this species is not obvious.

Sexual maturity in *T. rostratus* is attained three months after weaning and the disappearance of animals from regularly-trapped areas suggests that longevity rarely exceeds 12 months (Wooller *et al.*, 1981). If this is the case, females may only produce two or three litters in a lifetime. Litter size varies between one and four (Table 5.1), but litters of one and four are rarely found beyond the earliest stages of pouch life. Pruning of litters of four presumably relates to the energetic demands of suckling large litters. Litters of one may be eliminated in favour of replacing the litter, or alternatively, a single pouch young may not produce sufficient suckling stimulus to maintain lactation (Wooller *et al.*, 1981).

The limited data available for marsupial nectarivores (Table 5.1) suggest that the latter resemble one another in most life history traits. The most conspicuous differences appear to lie in the duration of the breeding season and in the age at first reproduction. The duration of the breeding season may be governed by the flowering phenologies in their habitats.

Fig. 5.1. Percentage of female *Tarsipes rostratus* with pouch young. From Wooller *et al.* (1981).

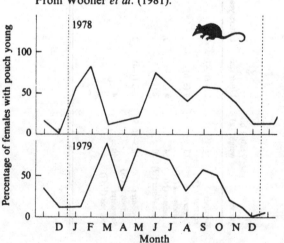

Frugivore/granivore

Although taxonomically aligned with the burramyid nectarivores, *Burramys parvus* produces only one litter annually (Dimpel & Calaby, 1972). This species inhabits scree slopes in alpine areas where snow covers the ground for up to six months of the year. Births occur during the second half of November, early in the snow-free period, and independent young are first caught in late December, so that the pouch life may be relatively brief. Young disperse between January and March when about two-thirds adult size (Gullan & Norris, 1981). Whereas the annual fecundity of *B. parvus* is less than that of the nectarivores, the lifetime fecundity may be similar. The little information we have on the natural longevity of the nectarivores suggests that they live for one or two years (Table 5.1), whereas female *B. parvus* are known to live to at least four years of age, first breeding in the year following birth (Mansergh & Walsh, 1983). This is long lived for a non-flying mammal of this size.

Exudate feeder/insectivore

The petaurids in this category are usually less fecund than the smaller nectarivorous possums, usually producing one or two litters annually, each of one or two young (Table 5.2). However, they are usually more long lived as they live to 4–5 years of age, with a reproductive life of 2–4 years.

We have most information for the life history of *Petaurus breviceps*. This species usually produces a single litter annually (Suckling, 1984). Births are confined to spring (September–November) so that lactation occurs when insects are common and prominent in the diet. The young attain independence in approximately 120 days (Smith, 1973). Survival of dependent pouch young was at least 94% and 56% in two years at Suckling's (1984) Gippsland study area, and in the first of these years at least 79% of young born in the study area were recruited into the trappable population. Growth of young is rapid (see Chapter 3) and their dispersal from the maternal range generally occurs prior to the next mating period (Suckling, 1984). Sexual maturity is attained by this time, and at least 80% of the females produce young in their first breeding season. Almost all females in other age classes produce young in each year. Suckling (1984) found that 81% of litters comprised two young and the remainder a single young, and that this proportion was consistent for females of all ages. Maximum natural longevity is of the order of five years for males and seven years for females, although most individuals live for less than five years.

Table 5.2. *Life history traits of exudate-feeding/insectivorous petaurids*

Species	♀ adult body weight (g)	Litter size	Annual fecundity	Months of births	Adult ♀♀ breeding/ year (%)	Age at weaning (months)	Mortality of dependent young (%)	Age at dispersal (months)	Age of ♀♀ at birth of first litter (months)	Natural longevity (years)	Source
Gymnobelideus leadbeateri	122–133	1.6 (1–2)	3.2	Apr.–June, Oct.–Dec.	≃ 100	3	0–100	11–26 (♂♂) 7–14 (♀♀)	12	—	A. P. Smith (1980, 1984)
Petaurus breviceps	115	1.8 (1–2)	1.7–2.5	Aug.–Dec.	> 95	3–4	56–94	10–12	8–15	5–7	Smith (1973); Suckling (1984)
P. australis	400–660	1	1	Aug.–Apr.	≃ 100	6–8	—	18–24	24	—	Fleay (1947); Russell (1980); S. Henry (personal communication)

Interspecific variation in the fecundity of exudate-feeding/insectivorous possums (Table 5.2) results from differences in the number of litters produced annually. *Gymnobelideus leadbeateri* (see Plate 5.1) produces two litters annually, one in April–June, the other in October–November. This may be related to the ability of this possum to maintain a higher year-round intake of protein by feeding on tree-crickets and spiders which are common under the decorticating bark of *Eucalyptus regnans* (A. P. Smith, 1980).

Plate 5.1. *Gymnobelideus leadbeateri* feeds on insects and plant exudates. (Peter Brown.)

The interval between conception and independence of young in
G. leadbeateri appears similar to that in *P. breviceps*, but growth is slower
(see Chapter 3) in spite of the year-round availability of arthropods. Even
so, dispersal and sexual maturity of females occurs at about the same age
as in *P. breviceps*. However, unlike *P. breviceps*, one-year-old animals may
be socially restrained from breeding by their inability to find breeding
territories (A. P. Smith, 1980). Males remain in their natal colony until they
are approximately 15 months of age and probably do not mate until after
dispersal.

Fungivore/omnivore

We suspect that a number of potoroines exploit this feeding niche
(Chapter 2). By considering them together we do not intend to imply that
they have identical diets. With the exception of *Hypsiprymnodon moschatus*, which produces two young (Johnson & Strahan, 1982), they are
monovular and polyoestrous. They resemble the majority of macropodines
in that pregnancy occupies almost the entire oestrous cycle. There is a postpartum oestrus which may result in conception. The resulting blastocyst
remains quiescent until suckling by the pouch young becomes intermittent
towards the time when it emerges from the pouch, or the pouch young is
lost. Birth and mating occur shortly after final emergence of the young
from the pouch, and the young-at-heel continues to suckle for at least
another month.

The potoroines appear to be more fecund than other macropods even
when body size is taken into account (cf. Table 5.3 and Table 5.8). This
is due to a shorter interval between births, which is largely the consequence
of a shorter pouch life and, in some potoroines, of a shorter gestation.

Within the Potoroinae there is a dichotomy between the three *Bettongia*
spp. and *Aepyprymnus rufescens* on the one hand, and *Potorous tridactylus*
on the other. The former have a relatively short gestation while
P. tridactylus has one of the longest gestations known among marsupials
(38 days), and this difference is reflected in the different interval between
births. It will be interesting to see whether other *Potorous* spp. conform to
this pattern. If so, an explanation of the cause of these differences may lie
in the different nature of their habitats. *Potorous* spp. tend to occur in dense
vegetation in areas of high and predictable rainfall (Guiler, 1958; Heinsohn,
1968; Seebeck, 1981) whereas the other potoroines occur in grassland,
woodland and heaths, where rainfall is usually lower and less predictable,
cover is less dense and the probability of fire greater (Ride & Tyndale-Biscoe,
1962; Sampson, 1971 in Parker, 1977; Johnson & Bradshaw, 1977;
Johnson, 1978).

Table 5.3. *Life history traits of marsupial fungivore/omnivores (Potoroinae)*

Species	♀ adult body weight (g)	Litter size	Annual fecundity	Months of births	Gestation (days)	Pouch life (days)	Time between births (days)	Age at weaning (days)	Age of ♀♀ at birth of first litter (months)	Source
Aepyprymnus rufescens	Up to 3500	1	3	All months	22–24[a]	114	119	155	10–11	Johnson & Bradshaw (1977); Johnson cited in Tyndale-Biscoe & Renfree (in press)
Bettongia lesueur	Up to 3500	1	3	All months	21.3	115	122	165	6–8	Tyndale-Biscoe (1968)
B. penicillata	1300	1	3	—	21	> 100	—	120–130	9–10	Sampson in Parker (1977)
B. gaimardii	1660	1	—	—	21.1	109	—	—	—	Rose (1978)
Potorous tridactylus	1020	1	2.5[b]	All months	38	126	140	—	12	Hughes (1962); Heinsohn (1968); Shaw & Rose (1979)
Hypsiprymnodon moschatus	571	2	—	—	(30–42)	147	—	—	> 12	Johnson & Strahan (1982)

[a] Moors (1975) gives a gestation of 21–30 days.
[b] Possibly 2 if anoestrus occurs in March–June.

The time taken by the potoroines to reach sexual maturity is similar to that taken by the small macropodines (cf. Table 5.3 and Table 5.8). Female *B. lesueur* have the shortest period of maturation (6–8 months) while female *P. tridactylus*, which are the least fecund potoroines, have the longest period of maturation (12 months). The small amount of information available suggests that births occur in all months. There is no information on natural longevity or length of reproductive life.

Arboreal browsing herbivore

Three families, the Petauridae, the Phalangeridae and the Phascolarctidae are represented in this category. The small pseudocheirids in the Petauridae are all browsing herbivores and are at least as fecund as the exudate-feeding/insectivorous petaurids whereas the larger phalangerids and the koala are less fecund (Table 5.4). The phalangerids and the koala usually produce only a single young per year and tend to mature later than the petaurids.

The smallest of the arboreal browsing herbivores occur in the genus *Pseudocheirus*. Small, one-year-old female *P. peregrinus* generally produce a single litter of two young, whereas large (> 0.8 kg) two- to four-year-old females are more likely to produce two litters annually of two to four young (Pahl, 1984) (Fig. 5.2). The tendency to produce a second litter appears to depend upon the body weight of the female over winter, and this may be related to the severity of weather in winter as well as to age. Populations

Fig. 5.2. Relationship between annual fecundity of *Pseudocheirus peregrinus* and female body weight in June for a population at Lysterfield, Victoria ($r = 0.44$, $n = 79$, $P < 0.0001$). Data from L. Pahl (personal communication).

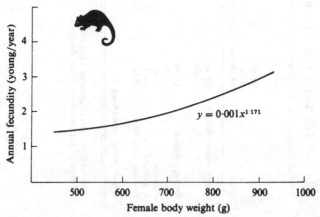

$$y = 0.001x^{1.171}$$

Table 5.4. *Life history traits of arboreal browsing herbivores*

Species	♀ adult body weight (g)	Litter size	Annual fecundity	Months of births	Adult females breeding/ year (%)	Age at weaning (months)	Mortality of dependent young (%)	Age at dispersal (months)	Mortality of dispersing young	Age of ♀♀ at birth of first litter (months)	Natural longevity (years)	Source
Pseudocheirus peregrinus	700–1000	2 (1–3)	1.8	Apr.–Nov.	90–100	6–7	—	8–12	—	12	5–6	Thomson & Owen (1964); Barnett, How & Humphreys (1982); L. Pahl (personal communication)
Petauroides volans	900–1700	1	0.68	Apr.–June	60–75	7–9	—	10–11	—	24	—	Smith (1969); Tyndale-Biscoe & Smith (1969); S. Henry (personal communication)
Trichosurus vulpecula	1500–3500	1	0.9–1.4	Mar.–May Sept.–Nov.	90	6–7	15	8–18	♂♂ high	12–24	< 15	Dunnet (1964); How (1978, 1981); Brockie, Bell & White (1981)
T. caninus	2500–4500	1	0.73	Mar.–May	80	8–9	—	18–36	—	24 (50%) 36 (50%)	♂♂12 ♀♀17	How (1976, 1978, 1981)
Phascolarctos cinereus	7900	1	0.65–0.70	Oct.–Mar.	65–70	11	[a]	12–18	—	24	12–15	R. W. Martin (personal communication)

[a] Negligible.

of *P. peregrinus* have been reported to differ in the proportion of females producing a second litter annually (e.g. 0% in Hughes, Thomson & Owen, 1965; 50% in How, 1978), but the cause of variation in these instances was not ascertained.

Most young of *P. peregrinus* are born in winter (June–July) so that lactation coincides with the spring flush in plant growth (Pahl, 1984). Sexual maturity is attained at approximately one year, when all surviving females usually reproduce. Maximum longevity is greater than six years in some populations (Barnett, How & Humphreys, 1982), but only four years in others (Pahl, 1984). This difference appears to be related to the abrasiveness of their diet and it's influence on tooth wear. Longevity is least in populations showing greatest rates of tooth wear, and emaciated individuals who disappear from populations have little useable tooth surface (L. Pahl & J. Gipps, personal communication).

The best-studied phalangerid is *Trichosurus vulpecula*, largely because of its economic importance in New Zealand where it was introduced in about 1840 (Pracy, 1974) and where it is a commercially important source of fur. However, it causes considerable damage to native forest and plantations of exotic conifers, and is also a vector for bovine tuberculosis.

T. vulpecula usually produces a single young annually, generally in autumn (April–June), but in a number of populations some females produce a second young in spring (September–October) (Dunnet, 1964; Smith, Brown & Frith, 1969). Young remain in the pouch for 170–190 days and are weaned between 200 and 250 days (Dunnet, 1964; Hocking, 1981). Males begin to disperse soon after weaning and most have dispersed by one year of age. Females, on the other hand, tend to remain close to the maternal range (Dunnet, 1964). Although females may breed at nine months of age (Pilton & Sharman, 1962), most mature in their second or third year (Smith *et al.*, 1969; Hocking, 1981). Maximum natural longevity is of the order of 12–15 years.

The demographic data available for *T. vulpecula* enable us to assess the range of tactical options open to this widespread generalist species. *T. vulpecula* is found in most woodland and forest communities in Australia, and has become established in many urban communities. Populations differ on the age at which individuals reach maturity, the incidence of breeding females, the rate of survival of dependent young and the probability of a second reproduction in spring. Increased spring breeding, early maturity and a high breeding success have been found to be associated in certain populations in New Zealand (Bell, 1981) and in the early stages of regenerating wet sclerophyll forest in Tasmania (Hocking,

1981). Hocking (1981) estimated that differences in these traits between populations in young (1–4 years) and old (60–80 years) forest regenerating after fire contributed to a 52% reduction in the intrinsic rate of increase *r* (Table 5.5). He suggested that variation in these traits resulted from changes in the quality of browse available to the possums in differently-aged stands, and supported this view with evidence of a decline in the nutritional quality of *Eucalyptus obliqua* foliage with age (Attwill, 1980). Others (e.g. Bamford, 1973; Boersma, 1974; Bell, 1981) have attributed this variability to changes in possum density and habitat quality. There is an inviting comparison to be made between the tactical options available to this species and those available to the seemingly more conservative habitat specialists *T. caninus* and *Petauroides volans*, but as yet there are too few studies for assessment of variability in their life history traits.

How (1978) drew attention to differences in the life histories of *T. vulpecula* and *Pseudocheirus peregrinus* on the one hand, and *T. caninus* and *Petauroides volans* on the other, and interpreted these differences in terms of *r* and K selection. *Trichosurus caninus* and *P. volans* were considered to be further towards the K end of the *r*-K spectrum than *T. vulpecula* and *P. peregrinus* because of their low frequency of litters, slow growth and late maturity (Table 5.4). He related this to the predictability of resources in the communities in which they occur. The wet sclerophyll

Table 5.5. *The intrinsic rate of increase of populations of* Trichosurus vulpecula *from different habitats*

Habitat	Median age at maturity (months)	Effective birth rate	Intrinsic rate of increase (*r*)	Source
Woodland (Canberra)	11	0.63	0.49	Dunnett (1964)
Regenerating wet sclerophyll forest				Hocking (1981)
1–4 years of age	17.4	0.42	0.34	
4–6 years of age	19.4	0.33	0.28	
8–16 years of age	26.3	0.29	0.21	
30–40 years of age	24.8	0.24	0.18	
60–80 years of age	29.7	0.24	0.16	
Podocarp – mixed broad-leaf forest (New Zealand)	24	0.22	0.17	Brockie *et al.* (1981)

After Hocking (1981).

forest and rainforest inhabited by *T. caninus* and the tall eucalypt forest inhabited by *P. volans* were seen as more predictable providers of resources than the variety of open forest, woodland and shrubland communities inhabited by the other species. However, the relative value for investment in young and juvenile survival in *T. caninus* and *T. vulpecula* are contrary to those expected from the *r-K* model, suggesting that the model does not apply in this instance. An alternative model, which incorporates variable mortality schedules and appears appropriate here, is the stochastic or bet-hedging model which predicts delayed maturity, iteroparity and low investment in young (Table 5.6). Unfortunately, there is no information on year-to-year and site-to-site variation in juvenile and adult mortality schedules, and we need this to test the appropriateness of this model.

Terrestrial herbivore

The reproductive strategies of the Macropodinae have attracted more attention than those of any equivalent marsupial taxon. However, reviews of marsupial or diprotodont strategies have, with a few exceptions (e.g. Tyndale-Biscoe, Hearn & Renfree, 1974: Newsome, 1975), been concerned with reproductive phenomena rather than with the ecological consequences of these phenomena, which is what concern us here.

There appear to be three ecologically distinct reproductive strategies among macropodines which differ in the timing of breeding and in the resulting fecundity. A group of species breed seasonally and have a fecundity of one young per year (Table 5.7). Species in this group tend to occur in temperate southern Australia where the quality and quantity of food resources are predictably good in spring and early summer, and poor in autumn and winter. They include the Tammar wallaby (*Macropus*

Table 5.6. *Predictions resulting from bet-hedging strategies with variable juvenile mortality*

	Environment	
Life history trait	Stable	Fluctuating
Maturity	Early	Late
Reproductive effort	Large	Small
Life span	Short	Long
Litter size	Large	Small
Number of litters	Few	Many

After Stearns (1976); Fleming (1979).

eugenii) and the Grey kangaroos (*M. giganteus* and *M. fuliginosus*). In contrast to these seasonal breeders, *M. agilis* from the tropical lowlands of northern Australia breeds year-round and has a fecundity of 1.8 young per year. Despite the strong monsoonal influence on its northern Australian habitat, and the consequent markedly wet and dry seasons, food may be locally scarce only during short-term flooding (Bolton, Newsome & Merchant, 1982). The third strategy is found in the Red kangaroo (*M. rufus*, see Plate 5.2) and Euro (*M. robustus*), which are opportunistic breeders and also have high potential fecundities (1.5–1.6 young per year). These species occur in arid Australia where droughts are frequent, unpredictable and of variable duration, and dramatically affect the quantity of food (Newsome, 1966).

While most large macropodines clearly employ one or other of these strategies, some of the small macropodines are facultative in their response, breeding seasonally in some populations and continuously in others. For example, the quokka (*Setonix brachyurus*) is confined to south-western Australia, including two offshore islands (Rottnest and Bald Islands). On the mainland, *S. brachyurus* is found in dense vegetation near creeks and swamps, which is likely to provide year-round nutritious food, and here breeds continuously (Shield, 1964). However, island animals breed seasonally. On Rottnest Island where *S. brachyurus* occurs in woodland, females come into oestrus in January and February and the young conceived at this time emerge from the pouch in spring, when there is a predictable flush of plant growth following reliable winter rainfall. Females

Table 5.7. *Reproductive strategies of macropodine marsupials (maximum annual fecundity in parentheses)*

Timing of breeding	Obligate breeders	Facultative breeders
Seasonal	*Macropus giganteus* (1) *M. fuliginosus* (1) *M. parryi* (1) *M. r. rufogriseus* (1) *M. eugenii* (1)	
		Setonix brachyurus (1.8) *M. parma* (1.3)
Continuous	*M. agilis* (1.8) *M. r. banksianus* (1.4) *Petrogale inornata* (1.7)	
Opportunistic	*M. rufus* (1.5) *M. robustus*	

then enter anoestrus and remain in this state until late summer (Sharman, 1965; Shield, 1968). Most females conceive postpartum, but a second offspring is produced only if the pouch young is lost. Animals in this population show an annual cycle in weight and haematological condition. Body weight and red cell parameters reach peak values in spring when young are weaned, and decline to their lowest values in late summer at the time of oestrus (Barker, 1974). Recovery of condition is associated with the spring flush of vegetation growth, and decline in condition with the effects of hot dry summers and overgrazing. Two pieces of evidence suggest that the pattern of reproduction found in *S. brachyurus* from Rottnest Island is associated with cycles in condition and food resources. Quokkas captured near a rubbish dump on Rottnest Island where food was continuously available carried young at all stages of development throughout the year (Shield & Woolley, 1963), and island animals lost traces of seasonal anoestrus over a two-year period when provided with food *ad libitum* (Shield, 1964).

Another example of facultative responses in small macropodines occurs in *M. parma*. In natural populations in New South Wales births can occur

Plate 5.2. *Macropus rufus*, a desert macropod. (Bruce Fuhrer.)

at any time, but most young are born between February and June (Maynes, 1977a). In contrast, a feral population on Kawau Island, New Zealand, showed seasonal breeding in the two years of observations prior to the development of pasture on the island, but thereafter bred continuously, only to return to seasonal breeding two years later when the pasture was severely overgrazed (Maynes, 1977b). In natural populations females may be sexually mature at 12 months of age, but on Kawau Island the youngest breeding females were estimated to be 19 months of age and most females did not breed before 2 years (Maynes, 1973a, 1976, 1977b).

The means by which seasonal reproduction is achieved differs among species showing this pattern. In the Tammar wallaby, 70–80% of the adult females on Kangaroo Island give birth in late January or early February and this is followed by a postpartum oestrus. A blastocyst produced as a result of conception at this time enters quiescence and remains in that state for the next 11 months. Young depart from the pouch in October and are weaned in November (Andrewartha & Barker, 1969). The blastocyst continues in quiescence until the summer solstice in December, when development recommences (Berger, 1966; Renfree & Tyndale-Biscoe, 1973). Thus these females suckle only one young at a time and produce only a single young each year. However, in some years a few females breed late so that the young emerge from the pouch in February. The mothers give birth again and so have two young sucking simultaneously, one of which is at foot. The predominant pattern of late summer births, pouch departure in spring and lactational and seasonal quiescence of the blastocyst is retained in captive animals and in a feral population in New Zealand (Berger, 1966; Renfree & Tyndale-Biscoe, 1973; Maynes, 1977b). A similar reproductive strategy is found in *M. r. rufogriseus* (Catt, 1977; Merchant & Calaby, 1981).

The Grey kangaroos and *M. parryi*, which is also a seasonal breeder, have an oestrous cycle which is four or more days longer than the period of gestation, and lactation suppresses postpartum ovulation. Pouch life is long (261–359 days) and as a consequence only one young is produced each year. In *M. giganteus* and *M. parryi* oestrus and ovulation may occur late in pouch life (*M. giganteus*, 112–243 days after birth (Kirkpatrick, 1965)) and may result in a quiescent blastocyst. The quiescent blastocyst reactivates as the preceding young begins to leave the pouch, and birth occurs a few days after final pouch exit (Kirkpatrick, 1965; Clark & Poole, 1967). Quiescent blastocysts have not been found in *M. fuliginosus* (Poole & Catling, 1974). The incidence of postpartum oestrus in *M. giganteus* is dependent upon the nutritional state of the female. At times of drought, few females ovulate during lactation, but in years of adequate plant growth as many as 85% of females may ovulate during this period (T. H. Kirkpatrick, cited by Poole, 1973).

The continuous breeder *M. agilis* and the opportunistic breeders *M. rufus* and *M. robustus* have similar reproductive traits (Bolton, Newsome & Merchant, 1982). They are characterised by a gestation period of approximately the same length as the oestrous cycle, a postpartum oestrus and ovulation from which a quiescent blastocyst is usually derived, and a pouch life of 235 days or less (Table 5.8). *M. agilis* (see Plate 5.3) shows high fertility in all seasons on farmland and bushland in the Northern Territory, despite a markedly seasonal climate (Bolton *et al.*, 1982). Over 90% of all adult females examined by Bolton *et al.* (1982) over a period of three years were fully reproductive and almost all of these were suckling a pouch young with a quiescent blastocyst *in utero*. Slightly more than one-third of these mothers also had an extra elongate teat indicating the presence of a young-at-heel. Seventy-two per cent of the small fraction of females in anoestrus were collected during the mid to late wet season, mostly from the bushland study site. The occurrence of anoestrus in these

Plate 5.3. *Macropus agilis* may breed year round. (Gordon Sanson.)

Table 5.8. *Life history traits of macropodine marsupials*

Species	♀ adult body weight (kg)	Litter size	Annual fecundity	Months of births	Gestation (days)	Pouch life (days)	Time between births (days)	Age at weaning (days)	Age of ♀♀ at birth of first litter (months)	Source
Macropus giganteus	3.5–32	1	1	Oct.–Mar.	36.4	297–359	362	>550	18	Kirkpatrick (1965); Poole (1975)
M. rufus	26.5	1	1.5	All months	33.2	235	236	365	14–22	Sharman (1963); Sharman & Calaby (1964)
M. fuliginosus	4.5–27.5	1	1	Oct.–Mar.	30.6	280–359	372	550	14	Poole (1975, 1976)
M. r. rufogriseus	14.0	1	1	Jan.–Aug.	30.5	230	365	400–500	11–22	Catt (1977); Merchant & Calaby (1981)
M. r. banksianus	13.8	1	1.4	All months	27.5	230	260	400–500	14–25	Merchant & Calaby (1981)
M. parryi	11.0	1	—	—	36.3	261–295	—	—	—	Calaby & Poole (1971); Poole & Catling (1974)
M. agilis	11.0	1	1.8	All months	29.4	207–235	200–215	288–328	10–17	Kirkpatrick & Johnson (1969); Merchant (1976)
M. eugenii	5.5	1	1	Jan.–Feb.	29.3	250	365	270	9–12	Berger (1966); Andrewartha & Barker (1969); Merchant (1979)
M. parma	3.2–4.8	1	1.3	All months	34.5	207–218	—	280	12–13	Maynes (1973a)
Wallabia bicolor	13.0	1	1.4	All months	35.5	256	—	—	8–10	Calaby & Poole (1971); Merchant in Tyndale-Biscoe & Renfree (in press)
Petrogale inornata	7.5	1	1.7	All months	31	204	210	286	18	Johnson (1979)
Peradorcas concinna	1.4	1	1	All months	30–32	180	335–385	300–360	12–24	G. Sanson (personal communication)
Setonix brachyurus	2.9	1	1[a]–1.8	All months[b]	27	185–195	185–365[a]	240	18	Sharman (1955a, b); Shield (1964)
Thylogale billardierii	3.9	1	1	Apr.–June	30	202	365	—	14–15	Rose & McCartney (1982)

[a] Rottnest Island.
[b] Rottnest Island, January–March.

females was attributed to sudden flooding of their dry season feeding sites causing a shortage of food. Births occurred all year round, although there was a peak in births at the bushland site at the time of transition from the wet to the dry season.

The strategy of the desert kangaroos is best exemplified by *M. rufus* which, in the presence of abundant green herbage, may produce an independent young every 240 days. Under these circumstances birth and postpartum oestrus follow within a day of final emergence of the young from the pouch. The young-at-heel continues to suckle from the same teat, and is weaned about 120 days later. The embryo derived from conception at postpartum oestrus develops into a blastocyst, but then remains quiescent until the pouch young begins to emerge about 190 days after birth. Final exit of this pouch young occurs some 45 days later and birth of the new young follows immediately (Newsome, 1965).

When there is abundant green herbage almost all adult female *M. rufus* will have one young-at-heel, another young in the pouch and a quiescent blastocyst. However, droughts are frequent within this species' range and these have a profound effect upon the quality and quantity of food, and consequently upon reproduction and survival. Newsome (1966) observed that in central Australia the abundance of green herbage decreases logarithmically during drought, and the level of reproduction in *M. rufus* is inversely related to the severity of the drought. On the best grazing land, one-half of the adult females in Newsome's (1966) study became anoestrous after a three- to five-month drought (Fig. 5.3). Many of these females continued to suckle young and roughly one-third of the young that were alive at the end of a drought were being suckled by anoestrous females (Newsome, 1965). Drought also impeded the growth rate of young, reducing their survival (Fig. 5.3) and delaying independence and sexual maturity. One-half of the pouch young died in a one-and-a-half- to two-month drought and none survived an eight-month drought. Loss of pouch young reactivates the quiescent blastocyst in reproductively-active females, and if drought persists, a succession of young are produced, each of which dies at about two months of age (Newsome, 1965) or later, but still while suckling (Frith & Sharman, 1964). Drought-breaking rainfall has an almost immediate effect on anoestrous females. Both Newsome (1964) and Sharman & Clark (1967) found that all females they examined were in proestrus or had ovulated within 14 days of drought-breaking rain. Many of these females were still suckling young born during the drought.

An interesting assessment of the effects of uncertain rainfall on population growth in *M. rufus* is provided by Newsome (1977). Newsome examined rainfall records for his central Australian study site for periods favourable for population growth during the years 1875–1974. With a knowledge of the responses of kangaroos, in terms of reproduction and survival, to drought (Fig. 5.3) and periods of pasture growth, and the monthly rainfall required to stimulate pasture growth, Newsome (1977) identified only 17 periods (17.7% of the time) that were adequate for 100% survival of pouch young (8 consecutive months of pasture growth) and a further 29 periods

Fig. 5.3. Relations in *Macropus rufus* between (*a*) the proportion of young which do not survive pouch life, (*b*) the proportion of females which are in anoestrus, and an index of aridity at a central Australian study site. The index of aridity represents the period during which drought is experienced, that is, when evaporation exceeds rainfall sufficient to cause seed germination and plant growth (for computation see Newsome, 1965).

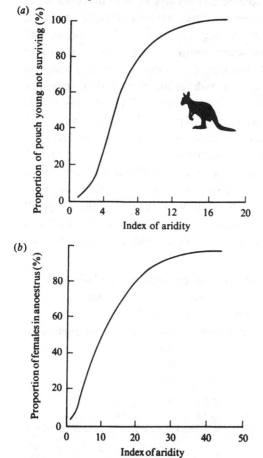

(31% of the time) that were adequate for 50% survival of pouch young. Nothing is known of the survival rates of these kangaroos in central Australia between nine months and three years of age, but from rainfall records Newsome (1977) calculated the probabilities of pasture growth occurring over periods of time sufficient for completion of pouch life, for the first year out of the pouch and for the additional period required to attain sexual maturity (Table 5.9). Newsome (1977) concluded that the low probability of occurrence of conditions suitable for population growth in successive years was responsible for the unusual inverted age structure of his study population (Fig. 5.4).

Parker (1977) and Low (1978) have argued that the marsupial mode of reproduction offers advantages in uncertain environments. They point out that in comparison with eutherians, marsupials exhibit low early parental investment. Gestation in marsupials demands little more than the resources committed to the oestrous cycle since it rarely exceeds that cycle in length, placentation is negligible and the young are minute at birth. Subsequent growth of the neonate is slow when compared with that of eutherian embryos *in utero*. Low (1978) also argued that marsupials may abandon pouch young with a lower cost and risk to the mother than accrue to eutherians faced with aborting young. Parker (1977) and Low (1978) concluded that marsupials may rescind reproduction when faced with conditions that are unfavourable for survival of young without incurring significant loss of the mother's reproductive value. However, even though the potential for litter abandonment may be a general feature of marsupials, it has only been demonstrated to be of any ecological significance in the desert kangaroos. Richardson (1975) drew attention to the range of strategies employed by the Macropodinae when he described the dichotomy between the reproductive traits of *M. rufus* and the similarly-sized

Table 5.9. *Probability of weather being adequate for survival of pouch young and juvenile* Macropus rufus *at a central Australian study site. Based upon climatic records for the years 1875–1974. Adequate weather for juvenile survival assumed to be the same as for pouch young survival*

Survival (%)	Pouch life	First year out of pouch	Sexual maturity
100	0.18	0.06	0.03
50	0.31	0.14	0.05

After Newsome (1977).

M. giganteus in terms of *r*- and K selection (see above discussion and Table 5.6 for the limitations to this approach). Thus even within the Macropodinae we have species which are clearly suited to predictable environments as well as those which are suited to unpredictable environments. As Morton *et al.* (1982) have noted in their critique of Parker (1977) and Low (1978), only 12.5% of Australian marsupials reside within the large arid region where unpredictability prevails, and too much emphasis has been focused on the reproductive modes of a few species, such as the desert kangaroos, to portray the variety of reproductive modes found in the marsupials accurately (see also Russell, 1982b). Nevertheless, the strategies employed by the desert kangaroos provide one of the most striking examples of mammalian life histories that are suited to environmental uncertainty.

In summary, the Macropodinae show three ecologically distinct reproductive strategies; one of these is characterised by seasonal breeding and low fecundity, another by year-round production of young and a relatively

Fig. 5.4. Estimated years of birth of *Macropus rufus* in samples collected between October 1959 and October 1962 in central Australia, and estimated periods of pasture growth (line). Solid histograms, animals aged on basis of molar eruption; dotted histograms, animals aged on basis of molar eruption and molar progression. From Newsome (1977).

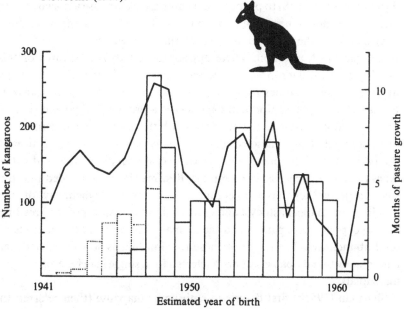

high fecundity, and the third by opportunistic breeding and a potentially high but variable fecundity. The availability of nutritionally adequate food appears to have a strong influence on the strategy adopted but explanations that are couched only in these terms or in terms of predictability of climate or resources (e.g. Bolton *et al.*, 1982) are likely to be too simplistic. In particular we need information on age-specific mortality before satisfactory explanations can be formulated. The placement of other macropodines within this framework requires more substantial information from natural populations than we have at present.

Embryonic diapause

There has been considerable speculation as to the ecological significance of embryonic diapause in marsupials, and for this reason we have dealt with it separately. Embryonic diapause, or quiescence, in marsupials appears to be confined to the Macropodidae, *Tarsipes rostratus* and probably a number of burramyid species (Renfree, 1981). Stasis in embryonic development is not confined to these taxa, but may be a general feature of marsupial intrauterine life. What is unique is the overlap of successive litters achieved by the presence of a blastocyst or blastocysts in a quiescent state in the uterus while the preceding litter is suckling in the pouch.

The widespread occurrence of embryonic stasis in marsupials led Tyndale-Biscoe (1968) to propose that embryonic diapause associated with lactation is not a new phenomenon but the extension of an existing adaptation of the embryo to intrauterine life which is derived from prolongation of gestation. There appear to be two key features of this development in macropods. First, the extension of gestation to occupy most if not all of an oestrous cycle, so that the female is in proestrus at or just prior to parturition and experiences a pre- or postpartum oestrus and ovulation. Second, the inhibition of the signal which terminates diapause so that the quiescent phase occupies a significant proportion of the pouch life of the preceding young. From a survey of this and other types of embryonic stasis in mammals, Renfree (1978) concluded that the sole property they have in common is to ensure synchrony of the development of the blastocyst and uterus with other aspects of the breeding cycle (i.e. pouch occupancy) and the animals' environment (i.e. periods of food abundance). The scope of these relationships is so substantial that it is unlikely that diapause has been selected for a single benefit, even among marsupials.

Sharman (1955b) first described embryonic diapause (then referred to as delayed implantation) in macropods and suggested that it facilitated the production of a second young in species which aggregate for only a short mating period. This view has not been sustained by observations of patterns of dispersion (Sharman, 1965). Sharman (1955b) also suggested that the diapausing embryo provided a means of replacing accidentally-lost pouch young without the need of further mating. Waring (1956) and Marshall (1967) elaborated on this advantage, pointing out that harassment by predators may cause large kangaroos to abandon pouch young and that the same harassment may cause females to become separated from males. Explanations such as this can only apply to the large plain-dwelling kangaroos (see Waring, 1956), or species where male fertility is highly seasonal, but again there is no evidence to support them. The view that the quiescent blastocyst provides an insurance against loss of the pouch young has been variously examined by Waring (1956), Ealey (1963), Shield & Woolley (1963), Sadlier (1965) and Marshall (1967). Shield & Woolley (1963) found little evidence of replacement when they examined the age distribution of pouch young of *S. brachyurus* on Rottnest Island. Almost all pouch young appeared to have been conceived shortly after females emerged from anoestrus in the summer. Further, they found that the capacity for replacement from blastocysts declined progressively from 60% in February, early in the breeding season, to 0% by August, as the pouch young began to emerge. This decline in capacity for replacement appears to be related to the nutritional status of females. Sadlier (1965) also questioned the value of the blastocyst as a means of replacing lost pouch young on the grounds that it would contribute little to the number of pouch young in the population. Using generous estimates for the frequency of females with quiescent blastocysts (60%) and the rate of loss of pouch young (10%) based on his observations on a population of *M. robustus*, Sadlier (1965) calculated that, for a cohort of 1000 females the number of females with pouch young would stabilise at 870 after three months, compared with 825 in a population without diapause. Although he viewed the difference of 5.5% as unimportant, it may well be sufficient for the character to become fixed in the population.

Another view is that the presence of a blastocyst reduces the time required to replace a lost pouch young. Ealey (1963) pointed out that for a kangaroo with a quiescent embryo it would take half the time needed by a kangaroo without one to replace a lost pouch young (Table 5.10). Although the time gained is only of the order of a month, this may increase the chance of the young completing pouch life during a period of weather suitable for its survival (see above).

Both Sadlier (1965) and Newsome (1965) have suggested that the main advantage of the quiescent blastocyst is extension of the period over which females have pouch young during a drought, so increasing the probability of possessing a pouch young at the time the drought breaks. During drought, pouch young are lost, and they are replaced with a reactivated blastocyst. Ultimately the female enters anoestrus even while suckling the remaining pouch young. Thirty-one per cent of the females Newsome (1965) observed emerging from anoestrus at the conclusion of a drought had suckled young throughout the drought. These young had an average age of 60 days, so reducing on average the period of favourable weather necessary for completion of pouch life to six months. While these benefits may be significant for macropods in unpredictable environments, they do not appear appropriate for the majority of macropods, which live in more predictable circumstances.

Another value of diapause in macropods can be seen in *M. eugenii* and *M. r. rufogriseus*. Here lactational and seasonal quiescence together provide the means by which reproduction is closely linked with predictable environmental events (see above). In *M. eugenii* the majority of females time the production of young so that weaning occurs during the period of greatest plant growth in spring (Andrewartha & Barker, 1969). The blastocyst remains in diapause until after the summer solstice, and birth occurs in late January or early February (Berger, 1966; Renfree & Tyndale-Biscoe, 1973). Curiously, this mode of producing a single young

Table 5.10. *Length of time taken by female macropodines with and without quiescent blastocysts to replace pouch young*

Species	Time from removal of pouch young to birth (days)		Source
	With quiescent blastocyst	Without quiescent blastocyst	
Macropus rufus	31.3	67.9	Sharman (1963); Calaby & Poole (1971)
M. agilis	26.5	59.2	Kirkpatrick & Johnson (1969); Merchant (1976)
M. eugenii	26.2	59.7	Merchant (1979)
Setonix brachyurus	25–28	53–54	Tyndale-Biscoe (1963)
M. parma	31.2	40.5–49.5	Maynes (1973a)
Petrogale inornata	29.0	62.0	Johnson (1979)

each year requires that the blastocyst survive for 11 months, and is surprisingly favoured over mating again.

The only general ecological explanation that has been provided for diapause in marsupials is that it prevents young of very different ages cohabiting in the pouch, and so avoids the danger of the larger young crushing the smaller young. Consistent with this view is an observation by Poole (1975) of two instances in which a second *M. fuliginosus* young entered the pouch before it was vacated by the larger neonate. (*M. fuliginosus* does not employ embryonic diapause.) In both cases the new neonate failed to survive cohabitation. Overlap of pouch occupancy by young of different ages occurs occasionally in *Trichosurus vulpecula*, a species without embryonic diapause, but we have no information on mortality under these circumstances which can be used to assess this hypothesis. If the above explanation is correct, then the acquisition of diapause by macropods can be seen as another aspect of the development of parental care which characterises this group. However, this begs the question as to the benefit these marsupials obtain from overlapping the development of successive litters.

With the exception of a few studies such as those of Shield & Woolley (1963), Sadlier (1965) and Newsome (1965), there have been few attempts to assess the ecological benefits accruing from embryonic diapause in marsupials. Even so, the data we have suggest a number of potential benefits which may differ between species. In some it appears to provide a means for spacing births so that certain life history stages coincide with predictable environmental events, in others it may ensure that females have a pouch young when droughts break and in yet others it may reduce the time taken to replace lost pouch young.

Social organisation

In Chapter 4 we described some aspects of the social behaviour of two dasyurids, *Antechinus stuartii* and *Sminthopsis crassicaudata*. Both species nest in groups, but whereas *A. stuartii* is highly gregarious and females are philopatric (Scotts, 1983), *S. crassicaudata* is often found nesting alone and is nomadic (Morton, 1978b). These differences appear to be related to the abundance and dispersion of food. Food is probably abundant and continuously available for *A. stuartii* and thinly scattered and unpredictable for *S. crassicaudata*. Other differences in social behaviour such as the marked differences in overt promiscuity and aggressiveness were seen as direct consequences of their life histories, and in particular of the opportunities that the males had for securing mates. These life histories were interpreted as being strongly shaped by the dispersion of food. The dispersion of food was therefore seen (though not proven) to be instrumental in shaping both the social organisations and life histories of these species.

Reviews of social organisation in birds and mammals (Trivers, 1972; Emlen & Oring, 1977; Clutton-Brock & Harvey, 1978) have concluded that the determinants of social systems, such as the dispersion of food and mates, generally act differentially upon the sexes. Trivers (1972) argued that this sex difference was a consequence of the disproportionate ability of the sexes to produce offspring. Female mammals, which incur the costs of gestation and lactation, cannot conceive and raise young at the same rate as males can sire them. The lifetime production of young by females is limited by the time and resources they can invest in this production rather than by the rate at which they can procure mates. The lifetime production of offspring by males is related to the number of eggs they can fertilise. Consequently, female strategies are more likely to be geared to the acquistion of food and the selection of the best possible mate and breeding site, whereas male strategies are geared to the acquisition of females. Intrasexual competition for these different resources is considered to play an important role in the evolution of mating systems.

Selection of the best possible breeding circumstances by the females will largely determine their dispersion and the degree to which the species is solitary or gregarious (Wittenberger, 1979). In this review we will pay particular attention to mating systems. Polygynous mating systems (Table 5.11) predominate among mammals because males have limited potential to care for young. The degree of polygyny will depend upon the ability of the males to monopolise females, and so will be dependent upon the dispersion of females. The potential for polygyny will be greatest where food resources are concentrated and clumped, and for monogamy where food resources are sparse and evenly scattered (Emlen & Oring, 1977). Monogamy may occur where resources are thinly and evenly scattered so that the distribution of females is such that a male can only defend the territory of a single female without reducing his reproductive success. Kleiman (1977) termed this form of monogamy as 'facultative' and distinguished it from 'obligate' monogamy where the male's assistance is required to rear his progeny successfully. This assistance may take various forms, from transportation of young to defence of territory (Kleiman, 1977, 1981).

One physical expression of mating systems is the degree of sexual dimorphism. In most mammals males are either larger than or of similar size to females (Ralls, 1976). Sexual dimorphism in favour of large body size in males has been associated with polygyny and intrasexual competition among males for females. Where males assist females in the care of young, sexual dimorphism tends to be greatly reduced. However, it is important to keep in mind the fact that intrasexual competition is not the only cause of size differences between sexes.

One other important life history trait which appears to be correlated with mating systems is the pattern of dispersal. In polygynous and promiscuous mammals juvenile males are the predominant dispersers (Greenwood, 1980). In monogamous mammals and birds juvenile females often disperse first and furthest (Greenwood, 1980), although often both sexes disperse (Dobson, 1982). Greenwood (1980) argues that in monogamous mating systems males will invest in resource defence to maximise their attractiveness to members of the other sex. This will favour philopatry of the resource defender and dispersal of females. This permits female choice of male resources and leads to avoidance of inbreeding. By contrast, a mate-defence mating system, in which males are primarily concerned with gaining access to and defending females, will favour male dispersal to increase access to females and also leads to avoidance of inbreeding.

Ganslosser (1982) and Russell (1984) provide substantial reviews of the social behaviour of marsupials and we will not repeat much of the detail they provide. There is relatively little information on the social organisation of the polyprotodont marsupials (see Chapter 4), probably because most are small, nocturnal and secretive. By contrast there is now some information on the social organisation of the larger possums, gliders and macropods and this forms the basis of our review.

Arboreal marsupials

Exudate feeder/insectivores. Despite similarities in diet, *Gymnobelideus leadbeateri*, *Petaurus breviceps* and *P. australis* show different patterns of social organisation. *G. leadbeateri* lives in colonies of up to nine individuals comprising one to three adult males, a single adult female and their progeny from one or two litters (A. P. Smith, 1980). Colonies of *G. leadbeateri* occupy territories of 1.3–1.8 ha and individuals within a colony share a nest located in a large hollow tree near the centre of the territory. These territories and nesting trees are evenly spaced throughout suitable forest (Fig. 5.5). The food resources of the possums are evenly scattered throughout the habitat and are therefore indefensible as localised resources. Evenly-scattered resources generally do not favour the development of territoriality (Brown, 1964), which is more energetically feasible

Table 5.11. *A classification of mating systems with some marsupial examples*

General classification	Spatial classification	Temporal classification
Monogamy: prolonged association and essentially exclusive mating between a male and a female	*Territorial monogamy*: monogamous pair shares common territory; *Gymnobelideus leadbeateri, Petauroides volans, Petaurus australis* (Victoria), *Trichosurus caninus*	*Serial monogamy*: individuals of both sexes usually pair with a new mate each year or breeding cycle
	Female-defence monogamy: each male defends access to a female instead of defending a territory	*Permanent monogamy*: mated pairs usually remain together for life, though sometimes they change mates after failed breeding attempts; *G. leadbeateri, P. volans, P. australis*
	Dominance-based monogamy: females maintain monogamous pair bonds within social groups by dominating more subordinate females	
Polygyny: prolonged association and essentially exclusive mating relationship between a male and two or more females at a time	*Territorial polygyny*: several females are paired with at least some territorial males; *Bettongia lesueur, Macropus rufogriseus, P. australis* (Queensland), *P. breviceps, Pseudocheirus peregrinus, Phascolarctos cinereus, T. vulpecula*	*Successive polygyny*: polygynous males acquire each of their mates in temporal succession
	Harem polygyny: a single male defends access to a social group of females	*Simultaneous polygyny*: polygynous males acquire all of their mates at same time; *M. rufogriseus, P. australis* (Qd), *P. breviceps, P. peregrinus, P. cinereus, T. vulpecula*

Successive polyandry: polyandrous females acquire each of their mates in temporal succession

Simultaneous polyandry: polyandrous females acquire all their mates at the same time

Polyandry: prolonged association and essentially exclusive mating relationship between a female and two or more males at a time

Territorial polyandry: several males are paired with at least some territorial females

Non-territorial polyandry: females desert first males and pair with new males elsewhere

Promiscuity: no prolonged association between sexes; multiple matings by members of at least one sex

Broadcast promiscuity: gametes are shed into the surrounding environmental medium so that sperm competition is prevalent

Overlap promiscuity: promiscuous matings occur between solitary individuals with overlapping home ranges or during brief visits by one sex to the home range or territory of the other; *Isoodon obesulus, Macrotis lagotis, Perameles nasuta, P. gunnii*

Arena promiscuity: males defend a display area or territory that is used exclusively or predominantly for attracting mates

Hierarchical promiscuity: males establish dominance hierarchies that affect their ability to inseminate females; *Antechinus stuartii, Macropus fuliginosus, M. giganteus, M. rufus, M. robustus*

Classification after Wittenberger (1979).

where resources are clumped and concentrated. However, A. P. Smith (1980) suggested that fixed area territorial defense may be possible in this instance because of the small size of the territory, the dense network of interlacing branches which characterise the habitat and facilitate movement, the speed and agility of the possums, and the sharing of the responsibility for defence by adults of the colony. The small territory size presumably reflects the richness of food resources in the wet, closed forest inhabited by this species.

A. P. Smith (1980) concluded that only one of the adult males in each *G. leadbeateri* colony is reproductive and that the species is monogamous

Fig. 5.5. (a) Distribution of *Gymnobelideus leadbeateri* nest sites (hollow circles) at Camberville, Victoria. Each nest site is occupied by an individual colony. Cross-hatching, forest habitat; black dots, unoccupied but potential nest trees. After A. P. Smith (1980). (b) Distribution of two colonies of *Petaurus breviceps* in roadside woodland at Rosedale, Victoria. Strong, broken lines, boundaries of the home ranges; hollow circles, sites of nest trees; cross-hatching, groves of *Acacia mearnsii*. After A. P. Smith (1980).

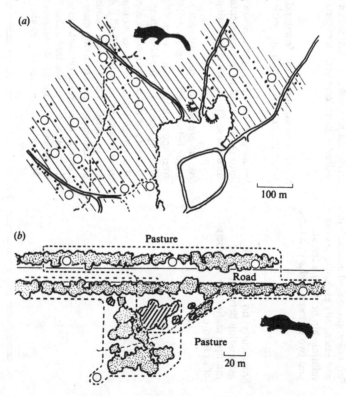

and not polyandrous. This conclusion was reached from observations of captive colonies in which members of a mated pair frequently performed mutual tail-licking behaviour, whereas other members of the colony never performed this behaviour. Mutual tail-licking behaviour was also observed in nature and presumably serves to maintain bonds between mates. Resident females attack intruding females and also maturing female offspring, which disperse before male offspring.

A. P. Smith (1980) also suggested that the assistance the reproductive male gives in territorial defence may be the basis for monogamy in *G. leadbeateri*, thereby insuring adequate food reserves for the female and their offspring. If Smith's interpretation is correct, then monogamy in *G. leadbeateri* is obligatory. The precise role and relationships of the remaining males in the colony is unknown, but they may be progeny as it is not uncommon for older male offspring to assist the dominant male in caring for the litter in instances of obligatory monogamy (Kleiman, 1977). The kinship of these males, their contribution to parental care and territorial defence and their success in supplanting or replacing the dominant male should be rewarding problems for further investigation.

The largest of the plant exudate/insect-feeding gliders, *P. australis*, also lives in small colonies. In tropical eucalypt forest at Atherton, northern Queensland, males were found associated with several adult females (Russell, 1980). In tall temperate eucalypt forest at Boola Boola, Victoria, colonies consist of a breeding pair and one or two offspring of up to 18 months of age (Henry, 1984a). These breeding pairs were observed to remain together for three years (Henry, 1984a). The difference in social organisation between these sites is probably related to the abundance of food resources, particularly arthropods, which are likely to be seasonally sparse and costly for a large glider to harvest in the forest at Boola Boola, and more continuously abundant and less costly to harvest in the forest at Atherton. These differences in resources are reflected in the colonies' home range size (40–60 ha at Boola Boola and approximately 2 ha at Atherton). At Boola Boola the home ranges of colonies appear to be largely exclusive (Henry & Craig, 1984). The very large home ranges at Boola Boola and the occurrence of polygyny elsewhere in the species range suggest that monogamy is facultative in this species. The distribution of resources at Boola Boola presumably results in widely dispersed females, thus reducing the probability that a male can defend the home range of more than one female. However, Henry (1984a) observed an independent young foraging and nesting alone with the adult male as well as the adult female, so that some paternal care of the young occurs in this species.

The third plant exudate/insect-feeding glider for which we have information is *P. breviceps*, which nests both singly and in colonies. In woodland at Rosedale, Victoria, A. P. Smith (1980) and Suckling (1980) observed that colonies had discrete membership and usually comprised five to nine individuals. The colonies varied considerably but commonly included three males and four females and their offspring in summer. Members of a colony tended to nest together in a tree hollow in winter, when huddling, nest sharing and torpor in groups reduce energetic costs (Fleming, 1980). These large groups broke up into smaller nesting groups in summer but individuals of a colony still shared the same home range. The home range of colonies showed little overlap (Fig. 5.5), and interactions between gliders from different colonies were rarely observed.

Henry (1984a) found no evidence of territoriality and considerable change in the membership of colonies in continuous forest at Boola Boola, Victoria. Here large groups, each consisting of 20–40 individuals and comprising six or so conventional colonies, occupied between 15 and 25 ha of forest. Individual colonies were of similar size and composition to those at Rosedale with the eight to ten individuals nesting together in winter breaking up into groups of three to five individuals in spring. Some colonies had stable membership for as long as eight or nine months, but others showed more or less continual change of membership. Large three- and four-year-old males changed colonies frequently, except during the breeding season in spring, when colony stability was greatest. The principal food resources of these gliders are scattered and make long-term defence of localised resources such as sap sites impractical. Therefore it seems likely that the differences in social organisation between the forest population at Boola Boola and the woodland population at Rosedale were related to the geometry of habitat. The narrow boundary between the colonies at Rosedale (Fig. 5.5) presumably allowed economic defence of territories which was not feasible in continuous forest.

The mating system used by *P. breviceps* is probably polygyny. One or two males were responsible for most or all of the mating in captive colonies studied by Schultze-Westrum (1965, 1969) and D. Hackett (in A. P. Smith, 1980), but the extent to which different males contribute to mating in natural colonies is unknown. A. P. Smith (1980) found that all females reproduced in natural colonies which contained fewer males than females.

Browsing herbivores. We have information on the social organisations of five arboreal browsing herbivores, which points to interesting differences between the species inhabiting wet and tall forest and the species inhabiting dry open forest and woodland.

Petauroides volans inhabits tall forest, and at the study site at Boola Boola, Victoria, occupied home ranges that were small (Fig. 5.6) and discrete between individuals of the same sex (Henry, 1984b) (Fig. 5.6). The home range of males usually coincided with the range of a single female, but in a few instances overlapped the ranges of two females. One male was

Fig. 5.6. (a) Home ranges of *Petauroides volans* at a study site in Boola Boola State Forest, Victoria. Arrows indicate males which courted females beyond their usual home range. Data from S. Henry (personal communication). (b) Home ranges of *Trichosurus vulpecula* in woodland near Canberra, Australian Capital Territory. After Dunnet (1964).

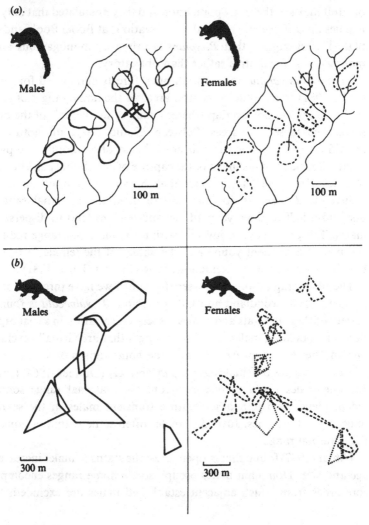

observed to partition time equally between two females, both of which produced young. Pairs nested together during the breeding season but broke up soon after the young emerged from the pouch. The pairs continued to associate, and came together 3–5 months prior to the next breeding season.

At Bondo State Forest, New South Wales, Tyndale-Biscoe & Smith (1969) found a tendency for gliders of the opposite sex to be more closely associated in their dispersion than animals of the same sex. They also observed that the adult sex ratio was strongly biased towards females (0.38) and that not all adult females bred in any year. There was a close correspondence between the number of females that bred and the number of adult males in the study population and they postulated that only paired females bred. Taken together, the observations at Boola Boola and Bondo State Forest suggest that *P. volans* is primarily monogamous and that monogamy is facultative rather than obligatory.

Trichosurus caninus inhabits tall open forests and closed forests. How (1978, 1981) found that the central areas of the home ranges of adults of the same sex did not overlap whereas pairs of individuals of the opposite sex shared these core areas. Between 75 and 100% of captures of an individual were confined to this core. These observations, the parity of adult sex ratios and instances of the capture of cohabiting pairs in the same trap, led How (1981) to conclude that this species is monogamous.

Juvenile *T. caninus* do not disperse from the maternal range until one-and-a-half to three years of age and females tend to disperse before males. The presence of a juvenile within the maternal range reduces the survival of dependent young and the ability of the female to reproduce, but the nature of these interactions is not known (How, 1981).

The remaining arboreal browsers for which we have information occur in open forest, woodland and tall shrubland. *T. vulpecula* is found in a variety of these habitats and shows associated variation in social organisation. In woodland, males have home ranges that are virtually exclusive of one another but each overlaps with the home ranges of several females. The home ranges of females overlap extensively (Dunnet, 1964) (Fig. 5.6). Juvenile males begin to disperse from the maternal range soon after independence and join other mature transient males in the search for unoccupied territories. Juvenile females often settle within or adjacent to the maternal range.

Winter (1977) found that in open forest the status of males increases with age and size. Dominant males occupy stable home ranges encompassing core areas from which adjacent established males are excluded. Young subordinate males disperse from the maternal range between 9 and 16 months of age and establish home ranges which are exclusive of those of other young males, but sometimes completely overlap the ranges of older males. As in woodland, male ranges overlap those of several females, and juvenile females often settle within or close to the maternal range. Winter (1977) found that the ranges of adult females also become more exclusive as the females age.

T. vulpecula nests in tree hollows (dens) which occur in the core areas and are the only part of the home range which are overtly defended from individuals of the same sex. These possums generally sleep alone, although on occasions a consorting male may share a den with an oestrous female. At the start of the breeding season old males may visit old females resident within their home ranges. Sometimes two males visit the same female; one of these is generally subordinate and usually in his first breeding season. Other males are often attracted to females on the night of mating but these are repelled by the consort. As the prevalence of male dispersal means that close kinship between the consorting males is unlikely, the benefits gained by the subordinate male require further investigation. Winter (1977) found that some females had no association with a consort, mated with one or more males and were not defended by any of the mating males. Males often adopted a new consort relationship after mating with the first female, often with females breeding for the first time. These come into oestrus later than old females. It seems that the predominant mating system is serial polygyny.

Pseudocheirus peregrinus also occurs in a variety of forest and woodland habitats as well as dense shrubland, nesting either in tree hollows or in dreys (nests of interwoven plant material). Dreys are often shared, commonly by an adult male and one or two females, one of which may be a daughter. These family groups commonly use three dreys, with individuals often nesting alone (Thomson & Owen, 1964; Pahl, 1984). The home ranges of family groups overlap, but the groups do not share dreys.

Russell (1984) concluded that this species is probably monogamous on the basis of the observation that 75% of nesting groups investigated by Thomson & Owen (1964) consisted of an adult male and an adult female. However, adult sex ratios vary from parity in populations at low density to female-biased ratios where densities are high (Pahl, 1984), and since all adult females breed (Thomson & Owen, 1964), the mating system may be monogamous or polygynous, depending upon density.

Phascolarctos cinereus inhabits open forests and woodlands. Eberhard (1972) and Gall (1980) considered that populations consist of individuals which feed in the same small group of trees for months on end (residents) and others which show no site fidelity (transients). These categories are not age or sex specific, although most independent immature males are transients. Eberhard (1972) found that only a small proportion of the occurrences of *P. cinereus* (14%) was in trees commonly used by others and on this basis concluded that the home ranges of residents were largely separate and represented territories.

A recent study of a population of marked animals by P. J. Mitchell (personal communication) has shown that the social organisation is probably more complex than this. Mitchell recognised three categories of males: large males which were dominant within their ranges, satellite males which occurred around the margins of these home ranges and males which were found close to and were tolerated by dominant males. During the breeding season in summer, Mitchell observed that dominant males became active at dusk and moved from tree to tree, apparently checking the status of females and fighting with and excluding satellite males from their vicinity. Dominants stayed close to and repeatedly mated with females presumed to be in oestrus. The home ranges of dominant males generally overlapped those of several females. These activities declined as the summer progressed and were not observed during the cold months. This species appears to be the most polygamous of these arboreal browsers and, consistent with this, shows the greatest sexual dimorphism.

Terrestrial browsing and grazing marsupials

The little that is known about the social organisation of the potoroines and small macropods suggests that most are solitary with male ranges overlapping those of several females. For example, in *Potorous tridactylus* the home ranges of males appear to be exclusive (Heinsohn, 1968) and the ranges of individual males overlap with those of several females (Kitchener, 1973). In captivity males fight in the vicinity of oestrous females (Heinsohn, 1968). Mating is promiscuous. A more stable polygynous mating system occurs in *Bettongia lesueur*, which nests in burrows shared by an adult male and one to several adult females. Males defend these burrows against intruders (Stodart, 1966).

Long-term associations of breeding pairs can occur in the small macropod *Setonix brachyurus* (Kitchener, 1973). On Rottnest Island, cover provided by small shrubs is limited and important in summer when food is poor and water scarce. Males defend shelter against other male intruders and share shelter with females. Kitchener (1973) found that males form a relatively stable linear dominance hierarchy in which older males are dominant.

The rock wallaby, *Petrogale inornata*, lives in stable harems comprising a territorial male and one or several females (W. Davies cited in Russell, 1984). The males defend rocky areas which are the daytime refuges of the colony, but different colonies share feeding areas.

Substantially more is known about the larger macropods, many of which are gregarious and form loose associations of individuals. Group size appears to be largest and membership most stable in the forest and woodland species such as *M. parryi*. Kaufmann (1974a) found that this species forms stable, discrete groups of up to fifty individuals which fragment randomly to forage. These groups are of mixed sex and age and range over 100 ha. The ranges of adjacent groups overlap and individuals from different groups mix freely at the boundaries. In contrast, *M. rufus* from arid shrublands and grasslands usually occurs in smaller groups whose composition changes continuously (Croft, 1981a). Large aggregations occur at patches of food or at isolated sources of water. *M. rufus* is generally sedentary where food is abundant, but is nomadic when food or water are scarce (Bailey, 1971; Dawson, Denny, Russell & Ellis, 1975; Newsome, 1975; Croft, 1981a). The other desert kangaroo, *M. robustus*, is sedentary rather than nomadic, but like *M. rufus* occurs in small, unstable groups (Croft, 1981b). It is found among rocky hills and makes use of caves for shelter during the heat of the day. Croft (1981b) considers that there is strong competition for caves, but found no evidence of defence of specific sites.

In the large kangaroos access to oestrous females is determined by size-based dominance hierarchies (Kaufmann, 1974b; Croft, 1981a, b; Johnson, 1983). Generally, the largest male in a group follows any female in proestrus and repels other males which attempt to mate with her. Sexual dimorphism is marked in these large macropods; this is consistent with their promiscuous mating system.

These dominance hierarchies among the males appear to influence the composition of the groups. Although the males show some tolerance of one another in groups, there is some tendency for avoidance or exclusion from groups. For example, Johnson (1983) found that male *M. rufus* were more often found alone or in mixed-sex groups and less often in single-sex groups than females. Further, large males tended to associate with lactating females on seasonally-preferred pastures. Johnson & Bayliss (1983) found that large males associated with females which had large pouch young or young-at-foot and were therefore likely to be near oestrus. In the seasonal breeder *M. fuliginosus*, males were found less often with other males and more often with females during the period when matings occur, but showed a tendency to live in groups during the remainder of the year. These groups may allow males to assess continuously their dominance status (Johnson, 1983).

In our introduction to social organisation we summarised the prevailing view among behavioural ecologists that the dispersion of female mammals is governed primarily by the distribution of food and to a lesser extent by shelter and predation, and that the dispersion of males is governed primarily by that of the females. In her review of marsupial social organisation, Russell (1984) concluded that most polyprotodont marsupials are solitary, and that this is a consequence of feeding upon insects which tend to be dispersed and require individual hunting. Predator avoidance is also best served by being secretive and solitary. This view is inconsistent with the evolution of complex social organisation in insectivorous Carnivora (Eisenberg, 1981; see Chapter 3). However, most diprotodont marsupials are more gregarious than the members of the polyprotodont taxa (Russell, 1984). The diprotodont marsupials mostly feed on plant parts or products which are frequently concentrated in their occurrence and therefore lend themselves to being harvested by groups of 'predators'. For large macropods group size and the stability of group membership depend upon the persistence of food resources, and whether individuals require specific shelters such as caves or rock piles. Where food resources are persistent, as in the coastal forest and woodlands of eastern Australia, group size is relatively large and group stability is high. Where resources are only temporarily abundant and are spatially heterogeneous, as in the central deserts of Australia, both group size and group stability are reduced. They may be further reduced where specific shelter sites such as caves are a limiting resource.

Group size tends to increase with body size and diurnality in macropods (Table 5.12). Jarman (1974) found a similar relationship between group size and body weight in the ecologically equivalent African antelope, and attributed the limitation on group size in small antelopes to their need to feed selectively in order to obtain nutritionally-rich items, which are scattered and so favour foraging alone rather than in groups. Many large antelopes are free from the need to feed selectively, and for them the advantages of living in groups may lie in the protection the group offers against predators. Predator avoidance may have little influence on grouping in marsupials. The only recent predators of the large kangaroos are the dingo and aboriginal man, but these have had only a short association (approximately 6000 years and 40000 years respectively). The extent of predation by extinct marsupial carnivores is unknown, but as far as the Thylacoleonids were concerned, was probably insignificant (Archer,

Table 5.12. *Group size in the Macropodidae*

Species	Body mass (kg)	Group size		Habitat	Source
		x[a]	Typical[b]		
Macropus rufus	17–85	2.2–2.6	2–5	Open shrubland, grassland	Russell (1979); Croft (1981a); Johnson (1983)
M. giganteus	3.5–66.0	2.1–6.4	4.4–8.8	Open woodland, forest	Kaufmann (1975); Taylor (1982)
M. fuliginosus	3.0–53.5	2.5	2–5	Open woodland, forest	Johnson (1983); Coulson in Croft (1982)
M. antilopinus	16–49	2.5–4.5	2	Tropical woodlands	Russell & Richardson (1971); Croft (1982)
M. robustus	6.2–46.5	1.4–2.2	2.6–3.55	Rocky hills	Russell & Richardson (1971); Taylor (1982)
M. agilis	9–27	—	1	Tropical open forest	P. M. Johnson (1980b)
M. parryi	—	6	—	Open forest	Kaufmann (1974b)
Wallabia bicolor	10.3–20.5	—	1	Thick undergrowth	R. Waters (personal communication)
Thylogale thetis	1.8–9.1	—	1	Dense forest	K. A. Johnson (1980b)
Potorous tridactylus	0.66–1.64	—	1	Thick undergrowth	P. G. Johnson in Strahan (1983)
Hypsiprymnodon moschatus	0.36–0.68	—	1	Rainforest	Johnson & Strahan (1982)

[a] Mean group size.
[b] Typical group size is average size of group in which species is typically found.

1981b). The life history stage most vulnerable to predation in these macropods is probably the young-at-foot, and both Croft (1981a) and Russell (1984) have argued that the high proportion of solitary females with young-at-foot in some species of macropods suggests that predation has had little significance in the development of grouping behaviour in the Macropodidae.

Contrary to the trend in macropods, group size in possums and gliders tends to decrease with increasing body size. In small gliders, such as *P. breviceps*, group size increases in winter and this behaviour has been shown to confer energetic advantages (Fleming, 1980). Even so, the distribution and abundance of food resources should influence the dispersion of females and this may account for differences in group size and group composition between *G. leadbeateri* and the similarly-sized *P. breviceps*. The concentration of food resources in the closed forest inhabited by *G. leadbeateri* is sufficient to allow fixed area territorial defence. This is not feasible for *P. breviceps* in open forest.

Territorial defence appears to occur only rarely among marsupials, and only as defence of small fixed areas and nest sites. Fixed area territorial defence is largely confined to arboreal marsupials inhabiting tall wet and closed forests, and we suspect that this is related to the year-round concentration of food and the ease of movement through the canopy.

Russell (1984) suggested that in marsupials the long period between litters and the limitation placed upon paternal care by raising of the young in a pouch in those species with a well-developed pouch, may constrain the formation of long-term bonds between breeding individuals and favour serial polygyny or promiscuity. However, the information we have on arboreal marsupials does not support this contention. Instead there is a dichotomy between those species inhabiting closed wet forests, which tend to be monogamous and have female progeny which disperse before male progeny, and those species inhabiting open forest and woodland, which are promiscuous or polygynous and show dispersal of male progeny before female progeny. These dispersal patterns conform to the predictions of Greenwood (1980) that female-biased dispersal occurs where species are monogamous and male-biased dispersal occurs where the species are polygynous.

Russell's (1984) observation that promiscuous and polygynous mating systems predominate among marsupials appears more accurate for the Macropodidae. Long-term bonding occurs in a few instances where shelter is limited, but under these circumstances the mating system is usually polygynous.

Caution must be applied in reaching any conclusion about the frequencies of different mating systems in marsupials. Studies of social organisation have been made on less than one-twentieth of the marsupial species, and most of those studies are of either conspicuous macropods or large gliders and possums from eastern Australia. In most instances we have information from only a single population of each species, and so have no indication of variability in social organisation with habitat. The most complex social organisation is found among the arboreal possums and gliders but our observations are restricted almost entirely to inhabitants of temperate eucalypt forest. The array of phalangers and petaurids in the tropical forest of northern Australia and New Guinea provide exciting subjects for future study.

6

Antechinus *as a paradigm in evolutionary ecology*

In recent years the study of ecology has undergone a rapid shift in approach. This has been associated with an increased emphasis on evolutionary principles in analysis of ecological relationships. Historically, ecology was biased towards descriptive studies, but now most synthetic ideas are generated from theoretical approaches, often couched in turgid equations which are incomprehensible to biologists who lack mathematical training. Unfortunately, the explosion of mathematical theory has now outstripped the ability of biologists to verify its main predictions empirically. The loose use of the results of theory which may be based on naive assumptions is something of which we should all be careful, and in Chapter 7 we review the growth of competition theory as an example of the inadequate integration of theoretical, experimental and descriptive aspects of science.

A further consequence of these developments has been a decline in the role played by mammals and birds in our understanding of theoretical ecology. Although the 'fathers' of evolutionary ecology, David Lack and Robert MacArthur, were both ornithologists, it has become increasingly apparent that invertebrates with simple life histories, simple population structure and short lives are superior empirical tools to most mammals and birds. In this chapter we discuss a fascinating mammalian exception to this pattern. The abrupt male mortality, synchronous breeding and monoestrous reproduction in *Antechinus* spp. overcomes many of the technical difficulties experienced by population ecologists. Animals may be aged precisely, the distinction between semelparity and iteroparity is clearly defined, and the maternity and survival of young is readily assessed. In the ensuing discussion we provide three instances where these unusual characters have

162

assisted with our understanding of general theory, and hope to leave the reader with the impression that there is potential for extrapolation of this approach.

Reproductive effort, stress and mortality

Central to life history theory is the relationship between reproductive effort and longevity (Calow, 1979), yet little is known of the mechanisms behind this relationship. Its expression should be most obvious where reproductive effort is intense and confined to a short period of the life history. Under these circumstances longevity should be markedly reduced. This expression is exemplified by mortalities which are associated with bouts of reproduction and which can be dissociated from shortages of food and predation. Examples are the 'spring declines' in numbers which characterise the life histories of several small mammal species (Krebs & Boonstra, 1978). These declines result from loss of numbers through mortality and dispersal of a substantial proportion of the population when animals come into the breeding condition in late winter or spring. Although they have been attributed to aggressive interactions (Turner & Iverson, 1973; Krebs, Halpin & Smith, 1977), their precise causes are poorly understood (see Gipps, Taitt, Krebs & Dundjerski, 1980; Lee & Cockburn, 1984).

Post-mating male mortality

The relationship between reproductive effort and longevity is particularly evident in small dasyurids. For example, in *A. stuartii* males are overtly preoccupied with copulation (see Plate 6.1) during the short annual mating period and all die abruptly at its conclusion. In contrast, the reproductive effort of females (including gestation and lactation) is spread over time and they are more long lived. In other similarly-sized dasyurids, such as *Sminthopsis crassicaudata*, males show no overt preoccupation with mating and survive the six- to eight-month mating period. A comparison between these diverse strategies provides insight into the mechanisms associating reproductive effort and longevity.

In order to understand the mechanism behind the synchronous and predictable post-mating mortality of male *A. stuartii* it is instructive to begin with the proximate causes of death. Autopsies have revealed a variety of disease states which could contribute to death, all of which are manifestations of a persistent state of stress. The most obvious and consistent finding of autopsy has been haemorrhagic ulceration of the gastric mucosa with substantial quantities of blood in the lower intestine (Barker, Beveridge, Bradley & Lee, 1978; Bradley, McDonald & Lee, 1980; McDonald *et al.*, 1981). Most individuals examined also showed evidence of increased invasiveness of parasites and resurgence of pathogenic micro-organisms which had probably invaded the host earlier in life and had remained as latent infections. Both states were associated with evidence of suppression of the immune and inflammatory responses of the host. For example, one sample of *A. stuartii* showed a high incidence of necrosis of the liver and this was attributed to infection by the ubiquitous bacterium *Lysteria monocytogenes* (Barker *et al.*, 1978). Local pneumonitis was frequently found at sites of lungworm (*Plectostrongylus* sp.) infestation and sometimes obliterated up to 50% of the functional respiratory surface (Bradley *et al.*, 1980). In addition moribund animals frequently showed moderate anaemias associated with heavy infestations of erythrocytes by *Babesia* sp. (Cheal, Lee & Barnett, 1976). Uniformly, these males showed low white cell counts, neutrophilia and lymphopenia (Cheal *et al.*, 1976), lymphoid follicles which were greatly reduced in size (Barker *et al.*, 1978)

Plate 6.1. Copulation in *Antechinus stuartii*. Note the loss of fur from the soon-to-die male. (Dan Irby.)

and a marked reduction in the concentration of serum immunoglobulin and in immunological competence (Bradley *et al.*, 1980).

Gastrointestinal ulceration and suppression of the immune and inflammatory responses are manifestations of a persistent state of stress in mammals and are a consequence of persistently high levels of free corticosteroids in the plasma. Stress is used here to describe a set of responses observed in animals when subjected to a variety of noxious agents (stressors) such as cold, injury or social insult (Selye, 1936). These responses enable the animal to combat the deleterious effects of the stressor. Central to this response is stimulation of the pituitary–adrenal axis leading to the release of corticosteroids. While corticosteroids facilitate the combating of stressors, prolonged stimulation of the pituitary–adrenal axis may lead to the so-called 'diseases of adaptation' such as gastro-intestinal ulceration and suppression of the immune response.

'Diseases of adaptation' (and 'adrenal exhaustion') have been invoked as a component of the mechanism behind stress-related population crashes in rodents (see Christian, 1971). However, the endocrinological basis of stress is poorly understood by population ecologists and for this reason we have chosen to elaborate on this basis. Only a fraction of the corticosteroids circulating in the plasma are biologically active and capable of eliciting a stress response. This biologically active fraction is determined by three variables: the release of corticosteroids from the adrenal glands, the metabolic clearance rate and the high-affinity corticosteroid binding system of the blood plasma. High-affinity binding of corticosteroids in the plasma is achieved by a protein, the corticosteroid-binding globulin (CBG), and this renders the bound corticosteroid inactive. Binding to CBG protects the corticosteroids from loss through metabolism in the liver and buffers the concentration of corticosteroids in the blood (Tait & Burstein, 1964; Baird, Horton, Longcope & Tait, 1969). The biologically active fraction of the corticosteroids is that portion which is not bound to CBG.

At the time of mating in *A. stuartii* there is an increase in the total plasma corticosteroid concentration (Fig. 6.1) and a precipitous decrease in the CBG concentration of males. Consequently there is a more than four-fold increase in the biologically active corticosteroid concentration. The fall in the CBG concentration of males is in part related to the increase in the levels of plasma androgens, principally testosterone. This has been demonstrated experimentally. Castration leads to an increase in CBG, while subsequent administration of androgens (testosterone or dihydro-testosterone) leads to a dose-related depression of CBG. This causal relationship is evident in field animals where there is a strong inverse correlation between plasma androgen and CBG concentrations (Bradley *et al.*, 1980; McDonald *et al.*, 1981) (Fig. 6.2). It is interesting to note that both experimentally and accidentally castrated males survived in the field beyond the period of natural mortality (I. R. McDonald & A. K. Lee, unpublished observations).

Plasma androgen concentrations increase with age in males, especially in the months immediately preceding mating, and attain exceptionally high maximum values just prior to mortality (Fig. 6.2*b*). Since there is no high-affinity binding of androgens in the plasma of these dasyurids, all the androgens are biologically active (Bradley *et al.*, 1980; Sernia, Bradley & McDonald, 1980).

The relationship between plasma androgen and CBG concentrations changes during the mating period (Fig. 6.2) at the time when plasma total corticosteroid levels increase. The decline in CBG levels relative to

Fig. 6.1. Levels of Total (circles) and CBG-bound (triangles) corticosteroids in the plasma of male (*a*) and female (*b*) *Antechinus stuartii*. Stippled area, free corticosteroids; solid bar, mating period; arrow, male die-off. After Bradley *et al.* (1980).

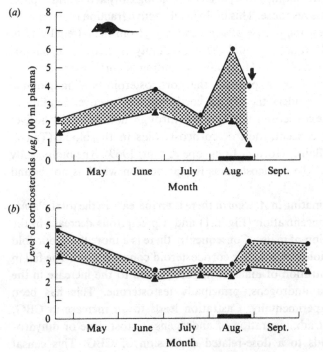

androgen levels potentiates the androgen effect, and appears to be due to increased release of adrenocorticotrophic hormone (ACTH) from the anterior pituitary; ACTH is responsible for the rise in total corticosteroids and also enhances the influence of exogenous testosterone on CBG concentrations in castrates (Bradley *et al.*, 1980; McDonald *et al.*, 1981). The basis of the increase in ACTH secretion during the mating period is still unresolved but could well be the intense reproductive effort of the males, including fighting and prolonged and frequent copulation (Chapter 4). Males deprived of the opportunity for this behaviour by isolation in

Fig. 6.2. (*a*) Relationship between CBG concentration and plasma testosterone concentration and (*b*) changes in plasma testosterone concentration with time of male *Antechinus stuartii*. Solid bar, mating period; arrow, male die-off. After Bradley *et al.* (1980).

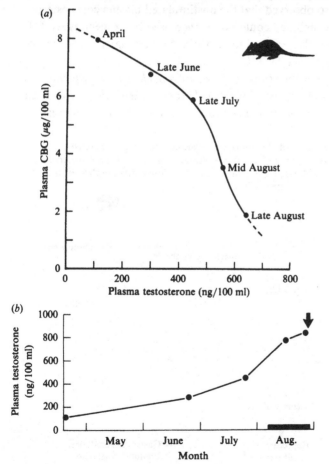

the laboratory survive well beyond the period of natural mortality. In contrast, males brought into the laboratory during the last week of the mating period die at the appropriate time.

There is evidence that the negative feedback which normally governs corticosteroid release from the adrenals fails in males at the time of mating. Under normal circumstances, any increase in the biologically active fraction is detected in the hypothalamus and this mediates a decrease in corticosteroid release by way of a reduction in ACTH release from the anterior pituitary. Failure of negative feedback has been demonstrated experimentally. I. R. McDonald & A. K. Lee (unpublished observations) found that at the time of mating there was no depression in the plasma total corticosteroid concentration following injection of a highly potent synthetic analogue, dexamethazone. This result is consistent with failure of feedback. They also observed that the unstimulated plasma corticosteroid levels were already high and could not be increased by exogenous ACTH. Therefore, the adrenals appeared to be driven to their maximum secretory capacity by endogeneous ACTH.

It appears then that the stress response in *Antechinus* spp. which leads to the post-mating mortality of males is 'programmed' and involves an increase in plasma androgen concentration and its effect in suppressing

Fig. 6.3. Synopsis of the endocrine changes leading to increased reproductive fitness and the post-mating mortality of male *Antechinus stuartii*. LH, luteinising hormone. Short arrows indicate direction of change; cross indicates failure.

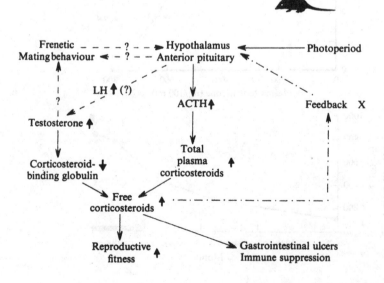

CBG, an increase in ACTH potentiating this effect, and a failure of the negative feedback system controlling corticosteroid release from the adrenal glands (Fig. 6.3). It therefore differs from the traditional view of stress in that this response is primarily promoted from within the animal rather than as an attempt to adjust to external stressors.

The probable significance of the stress response seen in male *Antechinus* is revealed by a comparison of the metabolic rates of free-living *A. stuartii* before and during the mating period. By using doubly-labelled water Nagy, Seymour, Lee & Braithwaite (1978) and Lee & Nagy (1984) showed that there was no substantial increase in metabolic rate at the time of mating. This would have been expected if the cost of mating had been added to the cost of living prior to mating. Since the cost of maintenance of general body functions, measured as resting metabolic rate, may increase by 17% in males at the time of mating (Cheal *et al.*, 1976), there may even be a reduction in the energy spent on other activities. These data suggest that the energy spent on mating behaviour must substitute for energy spent on activities engaged in prior to mating. There is indirect evidence that this may involve a reduction in feeding. Faeces are rich in chitin prior to mating but become watery and contain little chitin during mating (M. P. Scott, personal communication).

One of the principal benefits of stress is increased gluconeogenesis, the production of glucose from tissue protein, which increases the glucose available for energy production. Conceivably, increased gluconeogenesis in male *Antechinus* provides an alternative source of energy to that derived directly from the digestion of food and thereby allows substitution of mating behaviour for feeding. It seems therefore that the stress response in males facilitates a greater commitment of time to mating than would otherwise be possible. Both the negative nitrogen balance and the sharp decline in the weight of mating males (Woollard, 1971) are consistent with the hypothesis that tissue protein is a significant source of energy.

Other mammals and other life histories

The association between stress and post-mating mortality of males has been shown to occur in three dasyurids, *A. stuartii* (Bradley *et al.*, 1980), *A. swainsonii* and *A. flavipes* (McDonald *et al.*, 1981). A similar mortality occurs among males of the endemic rodent *Rattus fuscipes*, but in this species some adult males survive beyond their first breeding season (Warneke, 1971; Wood, 1972; Robinson, 1976). In southern Australian populations of this rodent, breeding occurs in summer and is largely confined to the months of December and January (Warneke, 1971;

Robinson, 1976; A. K. Lee, personal observations). Adult males relinquish the home ranges they have held during the spring and disperse during these months, and the number of transients caught on trapping grids increases markedly (Wood, 1972; Robinson, 1976). Aggression between males in dyad encounters in the laboratory also increases (Robinson, 1976). Analysis of the CBG concentration shows that it is lower than the plasma total corticosteroid concentration at this time in males, but not in females, which have high levels of survival. At other times of the year, the CBG concentration exceeds the plasma total corticosteroid concentration in most individuals of both sexes. Plasma androgens are also highest in males during the breeding period, when the relationship between the CBG and total corticosteroid concentrations favours an increase in the biologically active corticosteroid fraction (I. R. McDonald & A. K. Lee, unpublished observations). These data suggest that stress-related mortality may occur in males of this rodent during the breeding season, and that the phenomenon is not confined to dasyurids.

In the dasyurid *Sminthopsis crassicaudata*, in which some males survive through a six- to eight-month breeding season, the plasma CBG concentration of males remains substantially above the plasma total corticosteroid concentration throughout the breeding season so that there is no conspicuous increase in free corticosteroids. Although in a study by McDonald *et al.* (1981) plasma androgen concentrations early in the breeding season were higher than those measured in *A. swainsonii*, a species with a post-mating mortality of males, there was no correlation between plasma androgen and CBG concentrations. Understanding of the basis for this lack of correlation should clarify the distinctive mechanisms underlying these life history strategies.

Although analysis of the mechanistic basis of the life history strategies of mammals is in its infancy, it offers prospects for understanding certain of the problems which have perplexed ecologists working with small mammals. Not the least of these is the cause of 'spring declines' in many species of small mammals. These declines usually occur shortly after the onset of breeding in spring, affect males substantially more than females and may have a similar basis to the post-mating mortality of male *Antechinus* spp. (Lee & Cockburn, 1984).

Stress and population growth
The hypothesis that neurobehavioural–endocrine negative feedback may regulate the growth of populations of mammals was introduced to population ecologists by Christian (1950). An account of the development

of this hypothesis is given in Christian (1978). Briefly, Christian proposed that 'emotional and exertional stresses' associated with increased population density stimulate increased pituitary–adrenal and decreased pituitary–gonadal activity, and that these endocrine responses result in increased mortality and decreased reproduction. Initially, Christian identified the cause of increased mortality as exhaustion of the pituitary–adrenal axis as a result of prolonged stimulation, but has recently emphasised the importance of increased susceptibility to infection and parasitism, and diseases induced directly as a consequence of increased pituitary–adrenocortical stimulation (Christian, 1971; 1978).

Christian's hypothesis has not won universal acceptance, particularly in its application as a basis for explaining population cycles of microtine rodents and snowshoe hares. There is little evidence of stress-induced changes in mortality and reproduction in natural populations of mammals (see Krebs & Myers (1974) for this and other points of criticism). Most attempts to test the hypothesis have used some measure of adrenal function such as adrenal weight or total plasma corticosteroid concentration to quantify the endocrine status of animals, but the study on stress in *Antechinus* spp. indicates that these measures are far from adequate. In male *A. stuartii* adrenal weight and total plasma corticosteroid concentration increase at the time of mating, but similar increases occur in females, and they show little or no associated mortality. The significant difference between males and females lies in the concentration of biologically active corticosteroids. On average the level in males is two to three times the level in females (Bradley *et al.*, 1980) (Fig. 6.1).

Although the stress response in *Antechinus* spp. clearly differs from that invoked in the Christian hypothesis in that it is androgen driven rather than the product of external stimuli, the responses are sufficiently similar to suggest that the *Antechinus* study may provide a useful model for the design of future tests of Christian's hypothesis.

Geographic variation in litter size

In Chapter 3 we saw that the scope for variation in litter size is constrained by both body size and diet. While these constraints provide a useful framework for comparisons between higher taxa, they are of limited value in explaining the variation in offspring number within species or between closely-related species which do not differ in diet or size. Although increasing attention is being paid to differences in litter size between habitats (see e.g. Krohne, 1980), the best-documented examples of intraspecific variation in brood size are the tendency for avian clutches to increase with latitude (Lack, 1947, 1954) and mammalian litters to increase with both latitude (Lord, 1960) and altitude (Dunmire, 1960; Fleming & Rauscher, 1978; Bronson, 1979). A good example of latitudinal variation in litter size is provided by *Didelphis* spp. (Fleming, 1973; Tyndale-Biscoe & MacKenzie, 1976; O'Connell, 1979) (Fig. 4.7). Despite the generality of this phenomenon and the interest generated by David Lack's (1947, 1954) pioneering studies on variation in clutch size in birds, there is no consensus on its causes (Stearns, 1976; Ricklefs, 1980).

Lack (1947) had originally proposed that parents produce broods whose size corresponds to the maximum number of young they can nourish. He felt that birds at high latitudes would have more time to forage as summer daylength increased. This latter contention obviously generates different predictions for nocturnal mammals and diurnal birds, yet the two groups show a similar pattern of variation in brood size. In response to this and other problems, numerous modifications to Lack's hypothesis have been suggested (Ashmole, 1963; Cody, 1966; Owen, 1977; Hogstedt, 1980). A number of theoretical models have also been developed which suggest that brood size should be reduced below the most productive size (Stearns, 1976). Two classes with wide application can be recognised among these models:

(1) Those which suggest that as a result of a trade-off between fecundity (litter size) and adult mortality which results from increased reproductive effort, adults should limit the size of their brood in certain environments to extend their opportunities of breeding again (Charnov & Krebs, 1974). Earlier in this chapter we showed a relationship between adult mortality and reproductive effort in male *Antechinus*. While this and similar examples provide reasonable evidence for the proposed trade-off, specific models dealing with litter size have been criticised because they fail to predict what size smaller broods should be to ensure iteroparous reproduction (Ricklefs, 1977a).

(2) Those which predict trade-offs within the season the brood is produced between demands for resources for reproduction and demands for other functions, particularly the avoidance of predation (Cody, 1966; Skutch, 1967; Perrins, 1977). Ricklefs (1977b) presented evidence that predation is not sufficiently important to account for the magnitude of variation commonly observed along geographic gradients.

Several methodological difficulties have hindered investigation of these models. The genetic basis of variation in brood size has rarely been established (Krohne, 1980, 1981). Further, the influence of predation and the proposed trade-off between reduced clutch size and increased adult longevity have been difficult to quantify in field populations. While it is possible to distinguish between semelparous and iteroparous species, an index of the extent of iteroparity remains elusive and is confounded by difficulties in ageing wild individuals and by production of more than one brood within a season. A study of variation in brood size within semelparous species might be informative, but brood sizes in semelparous invertebrates and lower vertebrates and monocarpic plants are enormous and difficult to quantify.

Cockburn *et al.* (1983) suggested that *Antechinus* spp. offer an unusual opportunity to overcome these methodological problems. Because of the synchrony in life history events, the date of birth can be determined with unusual accuracy (compare with Pucek & Lowe, 1975). Males die after their first attempt at reproduction, and females produce one or two litters in their lifetime, depending on whether they survive through one or two breeding seasons. All young spend a period of 4–6 weeks when they are obligatorily attached to the teats of the mother. Although the number of young born may exceed the number of teats (Woolley, 1966b), the number of young born which can be weaned is equal to the number of teats. As teats are generally saturated with young (see Plate 6.2), teat number is a highly accurate index of litter size in any population (Fig. 4.4). This allows prediction of litter size in adult females which have yet to produce litters and in females captured after young release from the teats. As intrapopulation variance is very low or zero, teat number may be determined genetically.

Cockburn *et al.* (1983) mapped the distribution of teat number in populations of the four *Antechinus* spp. which have large geographic ranges in eastern Australia. All four species showed variation, from a minimum of six teats in *A. minimus*, *A. stuartii* and *A. swainsonii*, to a maximum of thirteen in some populations of *A. flavipes*. Although teat number in *A. minimus* was fixed within distinct subspecies, the variation found in the other three species did not conform to current taxonomic interpretation. It is consequently suitable for interpretation in an ecological context.

Teat number was lowest at low (northern) latitudes, and on exposed capes and promontories in the south (Figs. 6.4, 6.5, 6.6, 6.7), and highest in inland populations at high elevations. The range of variability in litter size is comparable with that seen in birds and other mammals which have been used in studies of macrogeographic variation (see e.g. Klomp, 1970).

The fecundity/survival trade-off hypothesis predicts that high female longevity would be restricted to populations with low teat number (low parental investment). The evidence for this association is equivocal (Cockburn *et al.*, 1983). Although females from eight-teated populations tend to produce two litters more often than females from ten-teated populations, other factors confound this association. The genus *Antechinus*

Plate 6.2. *Antechinus minimus* with young on each of its eight teats. (Chris Nicholls.)

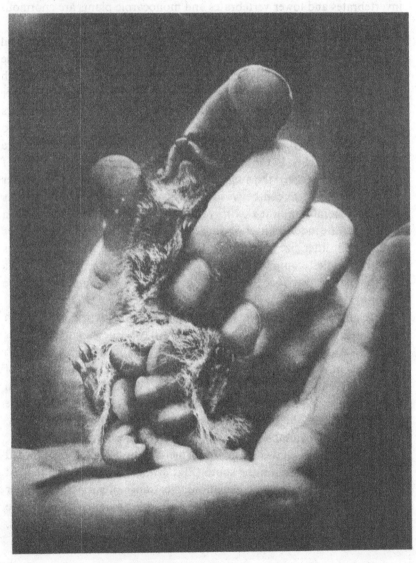

can be divided into two morphological guilds (Wakefield & Warneke, 1963, 1967) which reflect specialisation for either scansorial or soil-litter foraging (Braithwaite *et al.*, 1978). Scansorial species (*A. stuartii, A. flavipes*) are generally smaller than sympatric soil-litter foraging species (*A. swainsonii, A. minimus*). For reasons which remain uncertain, scansorial species are more often semelparous than sympatric soil-foraging species (see Cockburn *et al.* (1983) for a full discussion). The causes of the variability in female survival are of considerable interest and warrant further investigation, but it is dubious whether this variability explains variation in litter size.

Discussions of the influence of predation on brood size have concentrated

Fig. 6.4. Variation in the number of teats in *Antechinus stuartii* in eastern Australia. For data for Victoria (stippled area) see Figs. 6.5 and 6.6. Number of teats: filled triangles, 6; open circles, 7; filled circles, 8; open triangles, 9; squares, 10. After Cockburn *et al.* (1983).

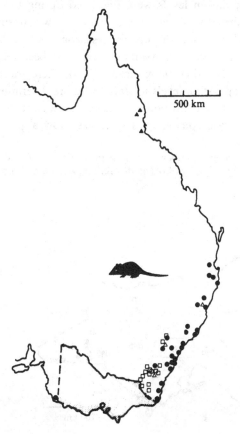

on characteristics peculiar to birds (Cody, 1966; Ricklefs, 1969; Perrins, 1977) and are consequently difficult to extend to *Antechinus*. In contrast to clutches of birds, litters of *Antechinus* are born rapidly, and the ability to feed all young simultaneously through lactation minimises the need to increase visits to the nest. We have no evidence which suggests that large litters are noisier than small ones (compare with Perrins, 1965). Although large litters may be more cumbersome in the final stages of pouch life, mortality of females during lactation occurs gradually and is not associated with any particular life history phase (Wood, 1970; A. K. Lee & R. W. Martin, unpublished observations). It is also hard to implicate predation in the low litter sizes exhibited by populations on the exposed capes and promontories of Victoria.

Lack's (1954) original proposal requires only slight modification to provide a plausible explanation of the observed pattern. Ashmole (1963) suggested that clutch size in birds is correlated with seasonal variation in carrying capacity. If population levels were regulated during the non-breeding season, litter size should reflect the extent to which surplus resources become available during the breeding season. Although this hypothesis has been largely neglected, Ashmole's view has been recently supported by Ricklefs (1980) and Grant & Grant (1980). There are strong reasons to believe that population levels in *Antechinus* are determined in the non-breeding season, when arthropod abundance is lowest (Lee *et al.*, 1977). In the comparatively aseasonal tropics, litter size is low, as predicted

Fig. 6.5. Variation in teat number in *A. stuartii* in Victoria. For data for (*a*) and (*b*) see Fig. 6.6. Symbols as in Fig. 6.4. After Cockburn *et al.* (1983).

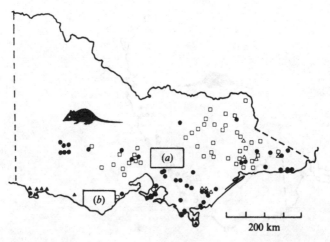

by the hypothesis. In New Guinean populations of *Antechinus*, females are polyoestrous and breed year-round, which reflects the low levels of seasonal variation (Dwyer, 1977; Chapter 4). In extremely seasonal highland areas (see e.g. Nix, 1976) in southern Australia, litter size is highest in all three species with broad geographical ranges. The capes and promontories of southern Australia may also have comparatively aseasonal climates and litter sizes are correspondingly low. These areas are constantly exposed to rain-bearing westerlies and have a highly predictable rainfall (Linforth, 1977). The proximity to the cold water of Bass Strait reduces

Fig. 6.6. Zones of transition of teat number in *A. stuartii*. (*a*) Marysville, Victoria; (*b*) Otway Ranges, Victoria. Symbols as in Fig. 6.4. After Cockburn *et al.* (1983).

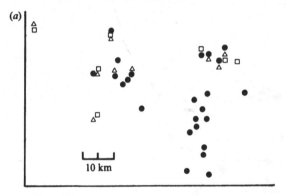

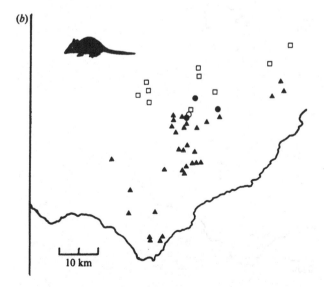

the amplitude of annual temperature fluctuations (Land Conservation Council, 1976; Linforth, 1977). In some localities along the coast, temperatures are never sufficiently low to retard plant growth, and moisture does not normally limit plant growth at any time during the year (Land Conservation Council, 1976). In constrast, at inland sites frequent

Fig. 6.7. Variation in teat number in *A. swainsonii*. (*a*) Eastern Australia; (*b*) Otway Ranges, Victoria. Symbols as in Fig. 6.4. After Cockburn *et al.* (1983).

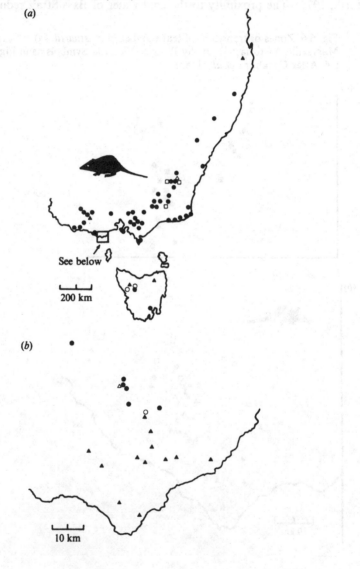

(*a*)

See below

200 km

(*b*)

10 km

frosts and low temperatures combine to retard growth during winter months (Land Conservation Council, 1976). Although studies of seasonal variation in arthropod abundance in eastern Australia are inadequate for us to contend with certainty that the availability of food for *Antechinus* corresponds with this gradient of seasonality, the connection between climatic seasonality and teat number is strong. Thus we contend that Ashmole's (1963) hypothesis is sufficient to account for macrogeographic variation in *Antechinus*, which may produce as many young as it can adequately nourish. This unique genus may be useful as a general paradigm for further investigation of variation in brood size in vertebrates.

Sex ratio

An important body of theory has arisen which suggests that there are fitness benefits for parents allocating resources to male and female offspring according to some optimum criterion. In particular, it is often assumed that the best parental strategy is to devote resources equally to the rearing of sons and daughters. This theory and assumption remain controversial. In the important text *The Evolution of Sex*, Maynard Smith (1978, p. 146) contended that:

> Work on the evolution of the sex ratio...forms one of the most satisfying threads in evolutionary theory. The ideas are elegant, and are often testable.

However, Williams (1979, p. 567), who had previously written the equally important *Sex and Evolution* (1975), has concluded:

> Evidence from vertebrates is unfavourable to...theory and supports, instead, a non-adaptive model, the purely random (Mendelian) determination of sex. The apparent absence of parental control of progeny sex ratio is a serious theoretical difficulty.

Clearly the resolution of this disagreement would be desirable.

Most theoretical treatments stem from Fisher's (1930) argument that a 1:1 sex ratio should be evolutionarily stable as there would otherwise be a frequency-dependent advantage to the rarer sex. Since Hamilton's (1967) demonstration that this result is dependent on the assumption of population-wide random mating, a number of theoretical studies have predicted that intrasexual competition or altruism should cause sex ratio bias (Werren & Charnov, 1978; Bulmer & Taylor, 1980; Taylor, 1981; Toro, 1982). Empirical support for this theory is restricted. Clark (1978) has argued that early male dispersal and female philopatry in the bushbaby *Galago crassicaudatus* leads to competition between female kin and a consequent advantage to mothers producing litters with a male-biased sex ratio. Criticisms of her study stress the inability to distinguish between alternative hypotheses (Hoogland, 1981; Masters, Centner & Caithness, 1982), and the generality of the result is weakened by the observation that male dispersal and female philopatry are widespread among polygynous mammals (Greenwood, 1980) while male-biased sex ratios are only occasionally reported (Clutton-Brock, Albon & Guinness, 1981). Stronger evidence for the hypothesis comes from the observation that the presence of resident female relatives depresses the reproductive success of adult female red deer, *Cervus elephas*. This also provides a possible explanation of the apparent bias in parental investment towards male offspring in this species (Clutton-Brock, Albon & Guinness, 1982). Clutton-Brock *et al.* (1982) invoked differential male dispersal as the cause of greater competition between mothers and their daughters than between mothers and their sons.

Cockburn *et al.* (1984a) were able to shed light on the importance of sex-biased dispersal for sex allocation by showing that sex ratios of pouch young in populations of *Antechinus* reflect the level of intrasexual competition, which varies with female longevity, while male-biased post-weaning dispersal is similar in all populations. Dispersal was investigated by amputating distinct combinations of digits from the forepaws of pouch young during the period of obligatory attachment to the teat. While

Fig. 6.8. Proportion of male and female offspring of *A. swainsonii* surviving in their mother's home range in the Otway Ranges, Victoria. Arrow indicates mating. After Cockburn, Scott & Scotts (1984).

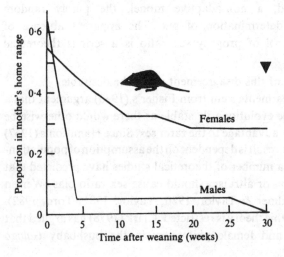

recovery rates at the time of weaning are similar for male and female offspring, all males disperse from the maternal home range within three weeks of weaning, and are replaced by an influx of immigrants. Some of these immigrants share nests with the resident mother and her daughters (Scotts, 1983; Cockburn *et al.*, 1984b). Female offspring show strong philopatry (Fig. 6.8). Although there may be an additional dispersal phase a few months before mating (Braithwaite, 1974), mothers (if alive) and at least one daughter continue to live in association beyond the dispersal phase.

Sex ratios may be measured while the young are attached to the teat, and are often female-biased in *A. stuartii* and male-biased in *A. swainsonii* (Cockburn *et al.*, 1984b). Populations from which samples are collected in more than one year show little year-to-year variation but high inter-population variance. This limited year-to-year variation reduces the applicability of the hypotheses we discussed in Chapter 3 which relate sex ratio to maternal condition. The extent of sex ratio bias in different populations is related directly to the degree of iteroparity among females (Fig. 6.9). This in turn is a measure of the potential competition between mothers and their philopatric daughters for any resource during the period

Fig. 6.9. Sex ratio of pouch young in *Antechinus* populations as a function of an index of iteroparity (proportion of parous females in the mating group). Filled squares, *A. swainsonii*; open squares, *A. flavipes*; circles, *A. stuartii*. After Cockburn *et al.* (1984a).

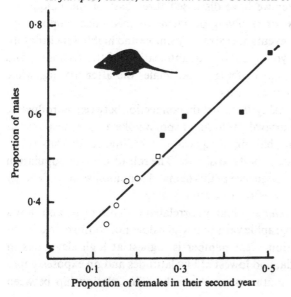

of reproduction (e.g. mates, food, nest sites). Other parent/offspring competition in *Antechinus* is likely to be trivial in comparison to mother/ daughter interactions. This is an obvious consequence of the non-overlapping generations of adult males and their offspring, and the post-weaning dispersal of all juvenile males from the maternal home range.

These results suggest that high levels of female intrasexual competition can influence sex ratio towards a male bias in *Antechinus*. Our study on *Antechinus* differs from previous analyses of local intrasexual competition as we distinguish between two causes of local differential intrasexual competition: the tendency of males to disperse further than daughters and variation in female longevity. All *Antechinus* spp. exhibit male-biased post-weaning dispersal, but sex ratio bias towards males is dependent on more prolonged mother/daughter association.

Antechinus are clearly excellent subjects for further investigation of sex ratio manipulation and dispersal patterns in mammals. However, it is important to remember that while simple life histories are both easier to analyse and conform more closely to the assumptions of mathematical theory than complex life histories, the simplicity of cohort structure might facilitate sex ratio variation in itself. Data which relate sex ratio to the probability of future reproduction are required before any generality can emerge.

Summary

In this chapter we have discussed three uses of the unique life history of *Antechinus* for resolving problems of theoretical importance. *Antechinus* life history events are extremely simple and highly synchronous, and access to pouch young enables quantification of the fecundity and reproductive performance of females. All males die after mating, while females breed in one or two years.

Analysis of male mortality illustrates the connection between reproductive effort and decreased survival. *Antechinus* spp. invoke a stress response as an adaptive means of mobilising energy without feeding, but this has a cost, an increase in their susceptibility to disease. The role of stress in population regulation may be a consequence of this demand for mobilisation of energy rather than a deleterious effect of overcrowding.

Teat number in *Antechinus* is highly correlated with litter size, and shows intraspecific macrogeographic variation which does not conform to current taxonomic interpretation. Teat number is highest at high elevations in south-eastern Australia, and lowest at low latitudes and on exposed capes and promontories in south-eastern Australia. The relationship between litter size and adult survival is equivocal, and predation pressure is unlikely to vary consistently across gradients of litter size. There is a close relationship between seasonality in resource availability and litter size, supporting Ashmole's (1973) modification to Lack's (1954) hypothesis as a parsimonious explanation of the evolution of variation in brood size. *Antechinus* may produce broods which correspond with the number of young they can nourish.

Antechinus spp. are unusual among outcrossed vertebrates in that they show sex ratio adjustment consistent with current theory. This adjustment may result either from the ease with which sex ratio variation can be recognised in this genus, or from the ability of selection to overcome Mendelian determination of sex in a genus with a simple and predictable life history.

The results discussed in this chapter are not exhaustive. Other areas made tractable by the unique life history are dispersal, population genetics, mother/offspring relationships, heterogeneity in reproductive performance and any system where the complexity generated by overlapping generations (see e.g. Charlesworth, 1980) precludes the empirical analysis of simple theory. We are convinced that this genus has a unique role to play in investigation of ecological theory.

7

Coevolution and community structure

Hitherto we have been concerned with the tactics adopted by individuals and the performance of populations, and we have demonstrated a number of applications of evolutionary theory to these levels of organisation. However, one of the most active areas of both theoretical and empirical ecological research concerns the interspecific interactions among individuals and populations. Such interactions fall within the ambit of community ecology or synecology. Most community properties are the sum of the properties of individuals and consequently reflect the separate adaptations of species. This causes difficulty in the application of evolutionary theory to explain the apparent structure and resilience exhibited by some communities with regard to phenomena such as species diversity (for an alternative view see Wilson (1980)). One type of individual selection which can increase the structure and stability of communities is evolutionary change in a trait of the individuals in one population in response to a trait exhibited by the individuals in a second population, followed by an evolutionary response by the second population to the change in the first. This process is called coevolution (Janzen, 1980), and may arise in a number of circumstances, depending on the nature of the effects of the interacting species on each other (Table 7.1).

Where two species interact to the benefit of both, the interaction is termed mutualism. Mutualism and other two-species interactions may arise without coevolution, but coevolution may enhance the relationship to the extent that the species become obligatorily interdependent (symbiosis). A more immediately obvious interaction is one where one species enhances its fitness to the detriment of another through predation or parasitism. Coevolutionary relationships between predators and prey leading to improvement of capture and avoidance efficiency are the best understood of two-species interactions. Another type of interaction is one where one species provides conditions which enhance (commensalism) or diminish (amensalism) the fitness of another without being affected by the other, but investigation of these relationships is fraught with difficulty and their importance remains uncertain (Begon & Mortimer, 1981). The last relationship of importance is competition, which occurs where two species cause a reduction in some aspect of each other's fitness. In many instances competition is easily discerned, but one of the most intense controversies within ecology concerns the extent to which properties of organisms reflect the consequences of coevolution between present or historical competitors.

In this chapter we concern ourselves with two of these types of interaction, mutualism and competition, which have recently been implicated in marsupial biology and which illustrate the difficulties and advantages of studying coevolution. We briefly review the history of study in each area, paying particular attention to recent controversy, and then discuss in detail marsupial examples pertinent to this context. We conclude with a brief discussion of the recent application of community theory to conservation biology and the design of nature reserves.

Mutualism

Mutualism, especially when developed into obligate symbiosis, is rare in nature (Briand & Yodzis, 1982). The best-known examples are those where one species lives within or in full association with the other, for example nitrogen-fixing *Rhizobium* bacteria in the roots of legumes and the fungi and algae which comprise lichens. Mutualisms where both species have an independent existence are less well understood. Mathematical models of this type of relationship suggest that the conditions for stable, obligate, coevolutionary association are quite restrictive, requiring stable, predictable environments. This may be why they are more prominent in tropical systems than elsewhere (May, 1981). Another problem is that

Table 7.1. *Forms of two-species interactions*

Effect of Species 2 on Species 1	Effect of Species 1 on the fitness of Species 2		
	Positive	Neutral	Negative
Positive	Mutualism		
Neutral	Commensalism		
Negative	Predation or parasitism	Amensalism	Competition

coevolutionary relationships between two species may cause the evolution of traits which are rather generalised in their effects (see e.g. Koptur, 1979; Janzen, 1980; Schemske, 1982), precluding analysis of the origins of the trait.

Marsupials as pollinators

One of the most fruitful systems used in the investigation of coevolution has been the interaction between flowering plants and animals which transfer pollen from flower to flower. The plants gain the benefit of pollination and provide nectar and pollen as an incentive to the pollinators. The morphology of flowers has evolved both to attract pollinators and to ensure efficient pollen transfer (Faegri & van der Pijl, 1979). While many plants may be pollinated by a variety of animals, or do not even require an animal vector for reproduction, other plants have evolved mechanisms that restrict pollination to small groups of animals or even individual species (see e.g. Powell & Mackie, 1966; Dressler, 1968; Ramirez, 1970). Groups of traits which appear to function for solicitation of a particular type of pollinator are called pollination syndromes, and characteristics which have proved useful for distinguishing these syndromes include the timing of anthesis and nectar secretion (Baker, 1960), the colour and morphology of the flowers (Kevan, 1978; Faegri & van der Pijl, 1979), and the volume, concentration and composition of the nectar (Baker & Baker, 1975). These syndromes have been best defined for groups such as hummingbirds, bees and moths with long proboscises (e.g. Sphingidae).

The efficacy of non-flying mammals as pollinators has attracted much recent debate, and some authors have contended that a pollination syndrome may be exhibited by plants specialised for this type of pollen transfer. Flower visitation by non-flying mammals seeking nectar and/or pollen is probably rather common, and has been reported among such unrelated taxa as Rodentia (Wiens *et al.*, 1984; Lumer, 1980), Primates (Coe & Isaac, 1965; Hladik, Charles-Dominique & Petter, 1980; Prance, 1980; Janson, Terborgh & Emmons, 1981), Carnivora (Lack, 1977; Charles-Dominique *et al.*, 1981; Janson *et al.*, 1981), Insectivora (Wiens *et al.*, 1984) and Macroscelidea (Wiens *et al.*, 1984). However, such visitation is particularly frequent among marsupials both in Australia (Turner, 1982) and in the New World (Charles-Dominique *et al.*, 1981; Janson *et al.*, 1981; Steiner, 1981). Evidence for pollination has mainly stemmed from these observations but for pollination to be effective it must both transfer pollen from anthers to conspecific stigmas and avoid destruction of the flowers, particularly the carpels. The relative efficiency of different pollinators can only be assessed by measuring the efficiency of pollen transfer and its frequency in time (Primack & Silander, 1975). Hopper & Burbidge (1982) took advantage of the diurnal and crepuscular activity of the Honey possum (*Tarsipes rostratus*) to analyse transfer of *Banksia grandis* (Proteaceae) and *Eucalyptus angulosa* (Myrtaceae) pollen by both *T. rostratus* and the White-cheeked honeyeater (*Phylidonyris nigra*, Meliphagidae, Passeriformes) in a coastal heathland in Western Australia. While both species carried pollen of both plants (Table 7.2), *T. rostratus* fed more slowly on *B. grandis*, were more sedentary and preened pollen from their bodies. These parameters may reduce both pollination rates and outcrossing potential, and have not been considered in other studies of non-flying mammal pollination. Preening behaviour is of particular interest, as it limits the value of measurements of pollen load on trapped animals (see e.g. Wiens, Renfree & Wooller, 1979; Hopper, 1980).

Turner (1982, 1983) has recently reviewed the evidence that certain traits of flowers and inflorescences reflect pollination by non-flying mammals. She showed that the importance of most of the traits had not been investigated experimentally, and where appropriate experiments had been conducted the results often contradicted previous speculation. For example, Carpenter (1978) suggested that the incurving of the styles of some *Banksia* spp. reflects specialisation for pollination by mammals. Hopper (1980) compared visitation and pollen transfer by birds and mammals on both

Table 7.2. *Feeding behaviour and pollen loads of Honey possums* (Tarsipes rostratus) *and White-cheeked honeyeaters* (Phylidonyris nigra) *(mean ± S.E.)*

	Tarsipes	*Phylidonyris*
Banksia grandis		
No. of probes/inflorescence	11 ± 3	24 ± 6
Probe duration (s)	18.7 ± 1.9	2.5 ± 0.3
No. of preens/inflorescence	2.3 ± 0.3	0
% of animals carrying pollen	27	83
No. of pollen grains/slide	0.3 ± 0.1	24.3 ± 11.1
Eucalyptus angulosa		
Feeding rate (s/flower)	3.3 ± 0.3	2.5 ± 0.2
% of animals carrying pollen	64	94
No. of pollen grains/slide	15.1 ± 13.9	29.3 ± 9.0

After Hopper & Burbidge (1982).

hook- and straight-styled *Banksia* spp. in Western Australia, and showed that hook-styled forms transfer more pollen to both vertebrate vectors.

This result illustrates another problem identified by Turner (1982, 1983), that is the widespread tendency to propose the importance of a trait as adaptive for a specific group of pollinators (e.g. non-flying mammals) when it may be equally important to other vectors. We have summarised traits of plants adapted for pollination by vertebrates other than hummingbirds in Table 7.3, ignoring characters unsupported by quantitative results or influenced by the restricted focus adopted by many workers familiar only with the idiosyncrasies of their local flora. There are few characters which can reliably distinguish plants adapted for pollination by birds, bats and non-flying mammals. While bats and non-flying mammals are likely to predominate as pollinators of plants with nocturnal anthesis and nectar production, no character clearly distinguishes the two groups of mammals. Indeed, the argument of Sussman & Raven (1978) that non-flying mammal pollination systems are only developed in parts of the world in which flower-visiting and fruit-eating bats are absent or sporadic is trivial and circular. Many flowers visited by neotropical marsupials are also visited by bats (Charles-Dominique *et al.*, 1981; Janson *et al.*, 1981; Steiner, 1981). The only neotropical plant species for which there is any evidence of specialisation for pollination by non-flying mammals is a *Blakea* sp. (Melastomataceae) occurring in cloud forest in Costa Rica, where rodents appear to be the most important pollinators (Lumer, 1980).

Nectarivorous bats are less abundant in southern Africa and southern Australia, where mammalian pollination has been recorded. Wiens *et al.* (1984) have reviewed the evidence that some members of the Proteaceae in southern Africa and their rodent pollinators represent a coevolved system. They conclude that the mutualism is not an example of strong coevolutionary adaptation as in contrast to the strong modification to *Protea* flowers, there is little evidence of specialised adjustment by the mammalian pollinators.

Conversely, in Australia three genera of marsupials from two families show behavioural and anatomical adaptation for a diet of nectar and pollen (Chapter 2). The Proteaceae have attracted considerable attention because it is within this family that non-flying mammal pollination has developed in southern Africa (Rourke & Wiens, 1977; Wiens & Rourke, 1978), and some Australian genera are clearly adapted for pollination by vertebrates (Armstrong, 1979). For example, although *Banksia* spp. receive visits from insects and birds as well as mammals, Turner (1982) considered vertebrates to be the principal pollinators because (i) the

Table 7.3. *Characteristics of plants adapted for pollination by vertebrates other than hummingbirds*

	Birds	Bats	Non-flying mammals	Generalist
Timing of anthesis and nectar production	Diurnal	Nocturnal	Nocturnal[a]	Diurnal and nocturnal
Volume and concentration of nectar and pollen reward	High	High	High	High
Sugar composition	Hexose-rich	Unknown	Unknown	Unknown
Robust flowers	Yes	Yes	Yes	Yes
Tight clusters of flowers	Sometimes	Sometimes	Sometimes	Sometimes
Colouration	Bright	Dull or light tone	Dull or light tone	Uncertain
Odour	Not strong	Uncertain	Strong	Uncertain
Height of inflorescence	Variable	Upper strata	Variable; low where trees and shrubs are sparse	Variable
Position of inflorescence on branch	Exposed	Variable	Variable	Variable
Mechanical triggering of anthesis	Sometimes	Sometimes	Sometimes	Sometimes

[a] *Tarsipes rostratus*, the non-flying mammal most strongly specialised for a diet of nectar and pollen, is active during both the day and night (Hopper & Burbidge, 1982). Some diurnal primates may also act as pollinators (Prance, 1980; Janson, Terborgh & Emmons, 1981).

Plate 7.1. *Acrobates pygmaeus* feeding at a *Banksia* inflorescence. (Mike Fleming.)

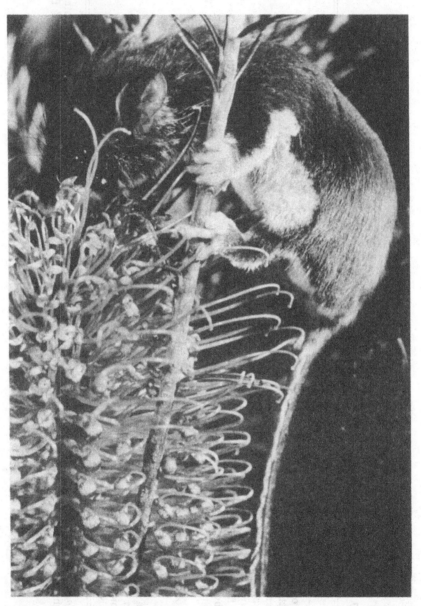

stigma–nectary distance is large, (ii) most banksias flower in winter when pollinating insects are neither abundant nor active and (iii) the large inflorescence structure and production of abundant nectar suggests encouragement of vertebrates as opposed to invertebrate vectors (see Plate 7.1 showing *Acrobates pygmaeus* feeding at a *Banksia* inflorescence). By contrast, the floral structure of most myrtaceous plants appears to be suited to both insect and vertebrate pollination.

The birds which visit banksias are principally honeyeaters and lorikeets (Loriidae: Psittaciformes). Some *Banksia* spp. appear to be specialised for honeyeater pollination (e.g. *B. coccinea*), and controversy has arisen as to whether specialist bird pollination is an apomorphic condition derived from a non-flying mammal pollination syndrome (Sussman & Raven, 1978; Ford, Paton & Forde, 1979), or vice versa (Rourke & Wiens, 1977; Carpenter, 1978). Turner (1982) has recently clarified this issue by pointing out that the association of marsupials, lorikeets and banksias is an ancient one, as relatives of all three groups are part of the Gondwanan biota. By contrast, ancestral honeyeaters probably entered Australia from Asia during the Miocene period. This led Turner (1982) to speculate that early evolution in *Banksia* was towards a generalist vertebrate pollination system (Table 7.4; Fig. 7.1). The subsequent radiation of the honeyeaters would have been facilitated by the presence of *Banksia*, and the success of honeyeaters precipitated the evolution in some banksias of diurnal anthesis and red inflorescences which appear to encourage bird pollination. Both

Table 7.4. *Traits of* Banksia *spp. and the marsupials, lorikeets and honeyeaters which pollinate them*

Pollinators	*Banksia*
Large body size	Flowers tightly clustered and strongly attached to stems
Arboreal and scansorial foragers	Inflorescence exposed and well above ground
Scansorial and terrestrial foragers	Inflorescence hidden and close to ground
Marsupial sense of smell well developed	Strong, sweet odour
Nectar feeders	Abundant nectar production
Pollen feeders (marsupials and lorikeets)	Abundant pollen production
Vertebrates active all year	Mainly winter flowering

After Turner (1982).

marsupials and lorikeets may digest pollen (Turner, 1982) while honeyeaters cannot (Paton, 1981). Turner (1982) argues that this is indicative of different periods of coevolutionary association.

If we accept Turner's (1982) thesis, we must conclude that there is no evidence that there are, or have been, Australian plant species specialised for pollination by non-flying mammals. As *T. rostratus* would probably become extinct in the absence of year-round availability of suitable flowering plants (Wooller *et al.*, 1981), this is an interesting contrast to the southern African situation, where unilateral specialisation occurred among plants rather than their mammalian pollinators (Fig. 7.2).

Summary. A number of authors have recently proposed the existence of a non-flying mammal pollination syndrome. Marsupials commonly visit flowers, and include all the mammalian species specialised for a diet of nectar and pollen. The plants visited by marsupials are also visited by bats (in the neotropics), and birds (in Australia) and these plants appear to possess traits reflecting a more generalised pollination syndrome than has been realised. Further investigations of pollination by mammals need to be empirical and quantitative rather than speculative. Such studies should recognise the importance of measuring the frequency and distance of pollen transfer by different vectors, the role of preening in restricting

Fig. 7.1. Coevolution of *Banksia* spp. and their pollinators. After Turner (1982).

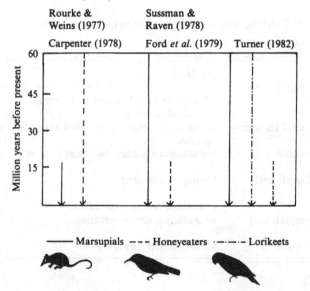

——— Marsupials – – – Honeyeaters ·—·—· Lorikeets

pollen transfer by mammals, and the generalised nature of many (?most) pollination systems. Until a more impressive array of quantitative data are available, prediction and synthesis will remain impossible.

Seed dispersal by marsupials

Like the pollination systems described in the previous section, it is easy to understand the potential for mutualism between a plant producing seeds and a vector which transports them. The plant can derive a benefit from colonisation of new habitat patches, avoidance of predation and reduction of local competition. Additional consequences include enhanced gene flow. Many plants have evolved fleshy fruits and arils which are accessible, conspicuous and easily digestible and so attract and encourage seed dispersers. Thus both types of mutualism involve provisioning of food by a plant in return for dispersal by an animal. Despite this basic similarity, seed dispersal differs from pollen dispersal in several fundamental respects which have been reviewed by Wheelwright & Orians (1982) (see Table 7.5). They point out that plants are less able to direct seed dispersal than pollen dispersal, and that the advantages of coevolution with a specific vector are much reduced. Flowering plants provide a reward at the time of pollen collection and deposition (see, however, Baker, 1976), while there is no similar incentive for seed dispersers to drop seeds in appropriate places (though this may occasionally happen (see e.g. Beattie & Culver, 1982)). This will minimise the differences between frugivores in their effectiveness as seed dispersers and reduce the potential for specialisation. Further, maintenance of a broad spectrum of dispersers may afford advantages by maximising fruit consumption and extending the range of habitats which can be colonised. It may even reduce the probability of extinction if a disperser becomes extinct. The consequences of specialisation

Fig. 7.2. Patterns of evolution in the South African and Australian Proteaceae and their mammalian pollinators.

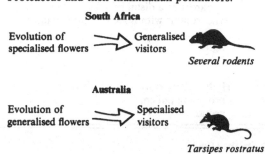

South Africa

Evolution of → Generalised
specialised flowers → visitors

Several rodents

Australia

Evolution of → Specialised
generalised flowers → visitors

Tarsipes rostratus

have been documented by Temple (1977) and Janzen & Martin (1982), who have suggested that the decline of some plants resulted from the extinction of their seed dispersers, the dodo on Mauritius, and the Pleistocene gompotheres (Proboscidea) in the neotropics.

In Australia the study of fruit/frugivore interactions has focused on birds (see e.g. Noble, 1975), and ants (Berg, 1975; Davidson & Morton, 1981a, b), and adaptation for dispersal by ants (myrmecochory) appears to be more common than elsewhere (Berg, 1975). There are no Australian or New Guinean marsupials known to be strongly frugivorous in their native habitat (see Chapter 2), but this may reflect the absence of good dietary studies of rainforest species. Clifford & Drake (1980) showed that in eastern Australia about 58% of plants in rainforests have indehiscent fleshy fruits of the type dispersed by vertebrates, while the comparable figures for heathland and dry sclerophyll forest are 12% and 20% respectively.

The only comprehensive study of the diet of didelphid marsupials showed that the genera *Caluromys*, *Didelphis*, *Marmosa* and *Philander* are at least partly frugivorous, while *Monodelphis* is not (Charles-Dominique et al., 1981; Atramentowicz, 1982). Of 138 plant species examined at Cabassou, in French Guyana, where the study was conducted, 58% were endozoochorous (seeds dispersed after being ingested), 14% were synzoochorous (seeds dispersed without being swallowed), 7% were anemochorous (seeds dispersed by the wind) and 21% were difficult to

Table 7.5. *General differences between pollen and seed dispersal*

	Pollen	Seed
Target site for dispersal	Stigma of conspecific flower	Site appropriate for germination
Characteristics of site as predictors of suitability	Distinctive colour, shape, etc., often apparent at a distance	Unpredictable
Temporal pattern of suitable sites	Synchronous with pollen dispersal	Unpredictable
Advantage for plant of diet specificity by vector	High	Low
Ability of plant to direct animal vectors	High: achieved by providing pollen, nectar	Low

After Wheelwright & Orians (1982).

classify. Large parrots and monkeys were absent, so potential dispersers through endozoochory included the marsupials, the kinkajou (*Potos flavus*, Carnivora), frugivorous bats and small frugivorous birds (Cotingidae, Thraupidae, Tyrannidae, Piperidae and Icteridae). Rodents (*Echimys* spp. and *Coendou prehensilis*) ate fruits destructively before they ripened, and did not contribute significantly to endozoochory. The range of vertebrates which facilitated endozoochory of a given plant species was restricted by a number of factors, of which seed size and hardness of the fruit coat were the most important. The role of vertebrates in dispersal of the common plants at Cabassou is summarised in Table 7.6. Birds mainly ate small fruits which have brightly coloured coats or arils and small seeds. Other small seeds were inaccessible to birds because they were unable to crack open the hard indehiscent shell or pod, and these seeds were most frequently dispersed by marsupials. While marsupials ate the pulp of larger species, they did not contribute to dispersal as the seed is not swallowed. The kinkajou was the major disperser of these larger seeds, except for the enormous seeds of palms, which are sticky and were dispersed through synzoochory by squirrels and agoutis. Seed size determines not only the range of potential dispersers but also the volume of seed reserves, and this ultimately determines the success of germination and the probability of seedlings surviving adverse conditions. Even in the humid tropical environment at Cabassou, the dry season represents a period of water deficit, and young seedlings are susceptible to dehydration. Because the onset of the wet season is very predictable, the optimum period for germination is the end of the dry season so that seedlings have the maximum time to develop before the following dry season. However, at Cabassou a fruiting peak at this time was much more pronounced among anemochorous plants than among zoochorous species, as many of the latter species have seeds which are dormant for some time prior to germination. This implies an additional selective factor influencing fruiting phenology.

Analysis of seasonal variation in the fruiting patterns of endozoochorous species enabled Charles-Dominique *et al.* (1981) to recognise four phenological patterns among the plants at Cabassou:

(1) Species with fruit production spread over an extended time period. These plants are early succession species, and their fruits have small seeds (Table 7.6).

(2) Species with synchronous, but irregular, cycles of fruit production. These trees have large seeds attractive to granivorous rodents, even prior to ripening. Massive, unpredictable fruiting cycles

Table 7.6. *The 26 most abundant plant species utilised by mammals at Cabassou, French Guyana, and the role of various vertebrates in the transport of seeds by endozoochory (E), synzoochory (S), consumption of pulp and rejection of the seed (R) or destruction of the seeds prior to ripening of the fruit (D). (Numbers in parentheses refer to the number of sympatric congeners)*

Seed size	Family	Species	Birds	Bats	Marsupials	Potos	Coendou Echimys	Small rodents	Phenology (see text)
< 20 mm	Euphorbiaceae	Pera bicolor	E	—	E	E	—	—	?
	Moraceae	Cecropia spp. (×2)	E	E	E	E	—	—	1
		Ficus sp.	E	E	E	E	—	—	4
		Ficus nymphaeifolia	—	E	E	E	—	—	4
	Melastomataceae	Henrietta succosa	—	—	E	—	—	—	1
		Bellucia grossularioides	—	—	E	—	—	—	1
	Araceae	Monstera sp.	—	—	E	—	—	—	1
	Passifloraceae	Passiflora glandulosa	—	—	E	—	—	—	1
	Annonaceae	Rollinia exsucca	—	—	R	E	—	—	3
	Burseraceae	Protium sp.	E	—	R	E	—	D	2
	Lauraceae	Ocotea spp. (×2)	E	—	R	E	—	—	3
		Ocotea or Nectandra sp.	E	—	R	E	—	D	3
Between 20 mm and 3000 mm	Boraginaceae	Cordia exaltata	—	—	R	?	—	—	3
	Simaroubaceae	Simarouba amara	—	—	R	E	—	—	3
	Rubiaceae	Guettarda macrantha	—	—	R	E	—	—	3
	Sapotaceae	Richardella macrophylla	—	—	R	E	—	D	2
	Myristicaceae	Virola spp. (×3)	—	—	R	E	—	D	3
	Mimosaceae	Inga spp. (×3)	—	—	R	E	—	D	2
> 3000 mm	Palmaceae	Attalea regia	—	—	R	—	S	—	3
		Astrocaryum vulgare	—	—	R	?	S	?	3

After Charles-Dominique *et al.* (1981).

reduce seed predation, but when isolated trees do not fruit synchronously with their conspecifics, they may have their crop totally destroyed by rodents.

(3) Species with synchronous and regular cycles of fruit production. These trees have large seeds, but chemical or physical deterrents protect the fruits until ripening.

(4) *Ficus* species where individuals fruit independently of each other, with a periodicity greater than or less than 12 months. These species attract a succession of frugivores.

Several species are dispersed principally by marsupials, and all have fruit production spread over an extended period (Fig. 7.3). Fruit at all stages of maturity are present throughout the year, and consequently ensure continuous exploitation by marsupials, and thus dispersal in both space and time. Larger animals may avoid these plants because the food supply is insufficient to warrant exploitation. This type of fruiting is restricted to plants with small seeds with limited reserves, which reduces the ability of seedlings to compete in mature forest. Consequently, most plants of this type are early succession forms. As Charles-Dominique *et al.* (1981) note that marsupials may be thirty to fifty times more abundant at early succession sites than in mature forests, marsupials may make an important contribution to tropical forest succession. As succession proceeds, invasion by plants with larger seeds would be facilitated by occasional forays from

Fig. 7.3. Fruiting phenology of plant species dispersed predominantly by marsupials at Cabassou, French Guyana. *Henrietta* and *Bellucia* spp. are closely related. After Charles-Dominique *et al.* (1981).

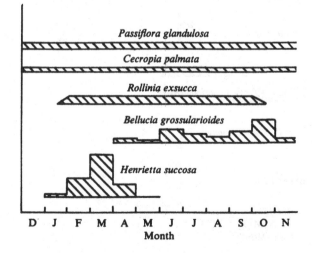

the larger frugivores. This would gradually lead to a floristic change, presumably to the detriment of the marsupials. The early successional status (< 80 years) of the forest at Cabassou may explain the low diversity of large frugivorous vertebrates.

Charles-Dominique *et al.* (1981) also suggest that the even phenological spread is adaptive to plants because it promotes persistence of frugivores. They extend this argument to explain the regular succession of fruiting among the large-seeded species dispersed by kinkajous. Frugivores require fruits dominated by both sugars and fats, and one species of each nutritional class appears to flower at any one time (Fig. 7.4). The

Fig. 7.4. Fruiting phenology of plant species with regular seasonal cycles at Cabassou, French Guyana. These species are dispersed by the kinkajou, *Potos flavus.* (*a*) Species with fruits rich in sugar; (*b*) species with fruits rich in fat. After Charles-Dominique *et al.* (1981).

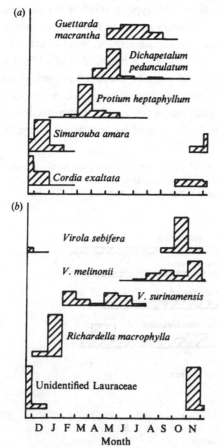

suggestion that this succession has evolved to facilitate year-round survival by frugivores clearly invokes a mode of group and community evolution not thought to be important by evolutionary biologists (Maynard Smith, 1975) as it postulates one species fruiting at a deleterious time to ensure the survival of other species. An alternative explanation for the even spread would be a gradual expansion of the time of year over which fruit is produced as a plant/frugivore relationship developed. Species or individuals with improved seed dormancy could fruit at times of the year other than the optimum period of germination, with the benefit of reduced competition for frugivores. Such competition is implied by the observation that fruit dispersal is reduced in regularly-fruiting species when fruiting happens to coincide with one of the species which fruits synchronously but irregularly (Charles-Dominique *et al.*, 1981). Available niche space would be fully exploited when plants fruited throughout the year, which would then ensure a large resident population of frugivores. However, it is important to regard the effect on the frugivores as a consequence rather than the cause of selection. Fruiting throughout the year may also be adaptive for early succession species because temporal spread of reproduction enhances the probability of successful colonisation (Comins, Hamilton & May, 1980).

Summary. Didelphid marsupials are important seed dispersers in tropical forest, and may influence the course of early forest succession. Plants dispersed exclusively by marsupials have fruiting spread throughout the year. Hypotheses which suggest that this pattern is designed to maximise the biomass and fidelity of the frugivore are group selectionist and inappropriate.

As a final comment, we should note that flowering and fruiting cycles are inextricably interrelated. Numerous accounts discuss the temporal pattern of either plant/pollinator or plant/frugivore interactions in isolation without adequate consideration of their direct connection. While some accounts analyse both effects (e.g. Milton, Windsor, Morrison & Estribi, 1982), further investigation of these complex relationships represents a potentially rewarding area for future research.

Sporocarpic mycorrhizal fungi

We have already commented on the frequency of mycophagy among the marsupials and other small mammals in Australia. Given the rampant speculation which surrounds study of other potential mutualisms, it is rather surprising that the possibility that sporocarps are designed to solicit consumption by mammals has not attracted greater attention.

Fungal spores are resistant to digestion (Martin, 1979), but there is some evidence that passage through the gut of a rodent may facilitate germination (Trappe & Maser, 1976). As sporocarps are often strongly odoriferous and attract a number of consumers it appears that some fungi have evolved a fruiting body which provides a reward in return for dispersal of spores. This relationship may be of importance in silviculture and horticulture (Maser *et al.*, 1978), and might be particularly important in Australia where soils are poor in nutrients (Specht, 1973) and many plant species rely on mycorrhizal associations because of the capacity of fungi to concentrate biologically important elements from extremely dilute substrates (Merrill & Cowling, 1966; Stark, 1972). Further investigation of these associations may be very rewarding.

Grasses and grazers: are they mutualistic?

While the potential for mutualism and coevolution between plants and the vertebrates which act as dispersers can be easily understood, plants and the animals which eat them do not immediately appear to be candidates for mutualism. Although there appears little doubt that there has been a close coevolutionary relationship between grasses (Poaceae) and the large mammals which graze them, the nature of this relationship has recently attracted controversy. During the late Miocene and early Pliocene periods there was a rapid radiation of grazing ungulates (particularly Bovidae and Cervidae) (see e.g. Janis, 1976; Webb, 1977; Cifelli, 1981), which appeared to occur at the same time as a major radiation of grass taxa and a substantial increase in the area of grassland (Stebbins, 1972). A similar scenario has been advanced to describe the transition among the Macropodidae from an ancestral browsing grade to an apomorphic grazing grade (Sanson, 1978). Grazing macropods show elaboration of the longitudinal link of the molars, reduction of the number of teeth in occlusion, broadening of the anterior cingulum and hardening of the enamel to accommodate an abrasive grass-based diet (Sanson, 1978, 1980). Molar progression also occurs in grazing macropods, obviating some of the problems associated with tooth wear (Sanson, 1978), although the origin of molar progression appears to be associated with selection on other aspects of dental morphology (Sanson, 1982). Macropods and ungulates which feed on grasses also show expansion of the foregut, which facilitates processing of a fibrous diet (Janis, 1976; Hume, 1978, 1982).

Grasses themselves possess traits which reflect a long association with grazers and include basal intercalary meristems that permit rapid compensatory growth, and variation in growth form and structure, and the composition of grasslands reflects the intensity of grazing (McNaughton, 1979; Mack & Thompson, 1982). While many of these traits have been interpreted as defence against herbivory, several authors, led by Owen & Wiegert (1976, 1981, 1982a, 1982b), have promoted the view that some aspects of the grass/grazer relationship can be viewed as resulting from mutualism instead of predation. To support this view they contend the following:

(1) Grasses, like many other plants, can be eaten without being killed (Owen, 1980) and some species of grasses disappear or are reduced in abundance in the absence of grazing (McNaughton, 1979).

(2) Because they contain fewer toxic secondary compounds than some other plants, grasses are more palatable to herbivores, 'soliciting' grazing to ensure defecation and urination in their vicinity to facilitate nutrient cycling (Owen & Wiegert, 1976; Stenseth, 1978; Petelle, 1982).

(3) The rapid compensatory growth and changes to growth form and nutrient allocation promote formation of extensive ramets (Owen & Wiegert, 1981), which may minimise the probability of extinction of any given grass genotype (Cook, 1979).

(4) Leaf abscission does not occur in most grasses, ensuring the availability of foliage to grazers in conditions unsuitable for photosynthesis. Grasses then require grazing (or burning?) to prevent accumulation of dead material (Owen & Wiegert, 1982a).

(5) The saliva of some mammals may stimulate grass growth (Detling, Dyer, Procter-Gregg & Winn, 1980; Dyer, 1980; Howe, Grant & Folse, 1982).

Benefits from herbivory have also occasionally been documented for other plants (Simberloff, Brown & Lowrie, 1978; van der Meijden & van der Waals-Kooi, 1979; Inouye, 1982).

These proposals have been criticised for a variety of reasons. Owen & Wiegert's (1981, p. 377) original claim that 'there is no question of grasses "defending" themselves from grazers' is not supported by comparisons of the palatability to mammals of different types of forage (Thompson & Uttley, 1982). Grasses may use silica particles in the blades and stems as a deterrent to grazing, necessitating the complex dental evolution described above (Herrera, 1982), and some chemical deterrents are also present (see e.g. Bergeron, 1980). The generality of salivary stimulation is unknown (Thompson & Uttley, 1982) and the conditions where nutrient cycling may operate to the benefit of grass and grazer are rather restrictive (Stenseth, 1978). As other traits of grasses can be treated with equal plausibility as a result of predator/prey or mutualism systems, we are sceptical of the general hypothesis proposed by Owen & Wiegert (1981).

Despite this scepticism, two benefits can be derived from this debate. First, it may be oversimplistic to regard all plant/herbivore interactions as deleterious to the plant, though this will be commonly true (Caughley & Lawton, 1981). Second, the debate again illustrates the futility of loose speculation concerning mutualism and other two-species systems without careful empirical research which assesses the fitness consequences of the interaction. For example, Inouye (1982) showed that damage by lepidopteran larvae to the central part of the basal rosette of the thistle, *Jurinea mollis* (Asteraceae), caused the production of multiple stalks, which elevated seed production. Other types of herbivory were deleterious to the plant, limiting the potential for development of general hypotheses concerning plant/herbivore interactions.

Competition
Competition, whereby species cause a mutual reduction of each other's fitness, is the opposite form of the species interactions which give mutualism. Unlike mutualism, interspecific competition is probably rather common, but the extent of its contribution to community structure is rather controversial. In order to analyse the nature of this controversy we feel it is necessary to review briefly the history of the study of competition, and its vocabulary.

Competition may arise through the joint use of a resource limiting to more than one species (exploitation) or may be developed to a state where direct antagonism between species occurs (interference). Identification of a limiting resource in the field can be difficult, and usually requires an experiment in which the resource is supplemented. The presence of competition should be easily confirmed with well-designed removal or addition experiments, though these are attempted only rarely. Some organisms, such as plants (Harper, 1977), and more-or-less sessile intertidal invertebrates (Connell, 1961), are more amenable to removal or addition experiments than are higher vertebrates. Laboratory habitats are even more suitable for manipulation, but often suffer from unrealistic simplicity. However, this simplicity leads to a paradox of considerable importance in ecological theory. While similar species often coexist in nature (e.g. *Eucalyptus* spp. in dry sclerophyll forest in eastern Australia or granivorous rodents in North American deserts), closely-related species rarely coexist in the laboratory (Gause, 1934), though the result of competition experiments in the laboratory is not always predictable (Park, 1954). The results of these laboratory competition experiments is summarised in the Competitive Exclusion Principle (Hardin, 1960), which states that two species cannot coexist on the same limiting resource. This principle also forms the basis of most mathematical analyses of competition (Pianka, 1981).

The niche

The coexistence of similar species in nature suggests that competition in the field is generally weaker than in laboratory environments, as species in complex environments have the opportunity to avoid competition by partitioning resources and habitats among themselves. In order to understand the nature of this partitioning it is necessary to introduce the concept of niche, which, despite some historical ambiguity, has proved to be a useful framework for the analysis of competition and resource use.

The awkward history of the niche has been well treated elsewhere (Vandermeer, 1972; Whittaker, Levin & Root, 1973; Whittaker & Levin, 1975; Hutchinson, 1978), and is recommended reading. We will content ourselves with provision of definitions which have wide current recognition, and a discussion of the problems which arise from them. The niche is a description of the components of the environment with which an organism, population or species interacts. A useful model of the niche is an *n*-dimensional hypervolume with each dimension describing a particular environmental component and consisting of the range of values of that component in which the fitness of the organism or species is positive (Hutchinson, 1957). Niche is thus defined from a species point of view, in contrast to habitat, which is a description of the environment and equally applicable to all the many species which live within it (e.g. desert, grassland), and role, which refers to the function of the species in the community (Hurlbert, 1981). Niche concepts are further refined by distinguishing between the fundamental or preinteractive niche, which is the niche which a species could potentially occupy in the absence of competition, and the realised or postinteractive niche, which differs from the fundamental niche according to the extent that interspecific competition reduces fitness to zero. Because of the difficulties in measuring fitness, which we have alluded to previously, descriptions of the niche have usually concentrated on resource use spectra (Pianka, 1981).

Begon & Mortimer (1981) have stated the Competitive Exclusion Principle in terms of niches: if there is no differentiation between the realised niches of two competing species or if such differentiation is precluded by limitations of the habitat, then one species will eliminate or exclude the other. We can generalise this contention by considering how resource use along a continuum can be altered to accommodate more species (Fig. 7.5). If there are no habitat limitations, increased resource diversity is possible. However, where habitat is limited, unless increased specialisation occurs, the niche overlap will increase to the point that niche parameters for two adjacent species will be almost identical. Biologists have been reluctant to elevate their hypotheses to the status of Principle and Law, and where this occurs it is often because the supposed Principle is only true because it embodies some circularity which renders it meaningless. The Competitive Exclusion Principle is certainly no exception, and has been branded as supremely unfalsifiable (Slobodkin, 1961; Simberloff & Boecklen, 1981), and as an 'unless it doesn't' theory by Hutchinson (1975). This is because within the context of the n-dimensional hyperspace model of Hutchinson (1957) it will always be possible to find some dimension which distinguishes two species.

Limiting similarity

The circularity inherent in the Competitive Exclusion Principle attracted considerable attention to three closely-related problems:

(1) Is there some degree of ecological similarity between species that cannot be exceeded without leading to extinction?

(2) Are natural communities saturated or packed with species whose similarities just approach this level?

(3) Does competition influence the number of species that can coexist in a community?

In his classic essay on this problem, Hutchinson (1959) placed these questions within an evolutionary framework for the first time (see Jackson (1981) for a review). Species diversity was of considerable interest at that time and in the ensuing decade because of controversy concerning the

Fig. 7.5. Schematic diagram showing how resource utilisation along a single dimension can be altered to accommodate more species. After Ricklefs (1979).

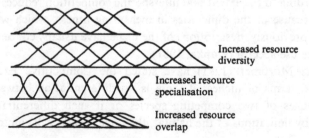

Increased resource diversity

Increased resource specialisation

Increased resource overlap

relationship between diversity (?complexity) and environmental stability. Hutchinson (1959) drew together the growing knowledge of patterns of diversity and the observation made by Brown & Wilson (1956) that certain species pairs show character displacement, or differ more in sympatry than in allopatry, and suggested that if the size of the trophic apparati of species differed by a sufficiently large ratio, then the species concerned would avoid competitive exclusion. He further suggested that the critical size ratio for some bird and mammal assemblages was 1.3.

Simberloff & Boecklen (1981) examined the influence of this idea, and suggested three derivative ideas that have become codified in the literature as hypotheses:

(1) There is a minimum size ratio compatible with coexistence of ecologically similar species (and this is often 1.3).

(2) Three or more ecologically similar coexisting species tend to have constant size ratios between species adjacent in a size ranking, though the constant factor may vary from site to site.

(3) The minimum size ratio and the constancy of size ratios are the results of interspecific competition and coevolution.

There have even been arguments which try to explain the physiological and evolutionary basis for a constant factor of around 1.3 (Maiorana, 1978).

A number of recent studies of marsupial species have used this perspective. Dickman (1982b) contended that the ratio of head–body lengths in populations of *Antechinus swainsonii* and *A. stuartii* in sympatry approximated 1.3, and the difference between the species facilitates niche segregation. Evidence for diet differences associated with body size are rather weak, with Hall (1980b) demonstrating a large overlap in prey size, though *A. swainsonii* takes more very large prey than does *A. stuartii*. (Fig. 7.6). Hall's results conform to Wilson's (1975) argument that the distribution function of prey sizes which may be eaten by a predator of given size will be asymmetrical. Some food will be too large for mechanical reasons, but the only limit to obtaining small food will be set by the amount of work done by large animals in gathering up small pieces of food, when large, more easily collected pieces are available. One experimental study using snails failed to show any body size/prey size relationship at all (Levinton, 1982).

In a more comprehensive analysis of body size patterns in dasyurid assemblages, Fox (1982b) showed that dasyurid species richness at any one site is low, and that assemblages do not exhibit close packing so that species are seldom near to limiting similarity (Fig. 7.7). He interpreted this in terms of lack of specialisation in dasyurids, and also argued that sexual dimorphism increased the niche width of each species, making less of the resource spectrum available to additional species. This latter hypothesis was originally invoked for birds (Selander, 1972), where the majority of species are monogamous (Greenwood, 1980). It seems less appropriate for polygynous mammals (Maynard Smith, 1982), as dimorphism in mammals is larger in polygynous and promiscuous mammals than in monogamous ones (Clutton-Brock, Harvey & Rudder, 1977). A further problem with Fox's (1982b) contention is his use of head–body length as a descriptor of potential niche width. The idea of minimum similarity arose from the contention that differences in feeding apparatus would facilitate species coexistence. If Van Dyck's (1980, 1982a) data are used to examine the extent of sexual dimorphism in *Antechinus* spp., it is clear that while there is a striking constancy in the dimorphism of head-body lengths, there is a reduced dimorphism in basicranium lengths, which are presumably related more closely to feeding success (Table 7.7). Sexual dimorphism is therefore unlikely to be related to feeding success, especially given the slight differences in prey size taken by different *Antechinus* spp. (Fig. 7.6). Sexual selection leading to dimorphism probably reflects the different life histories of the two sexes (Chapters 4 and 5).

Fig. 7.6. Frequency of prey sizes taken by sympatric populations of *Antechinus swainsonii* and *A. stuartii*. Stippled bars, *A. swainsonii*; open bars, *A. stuartii*. After Hall (1980b).

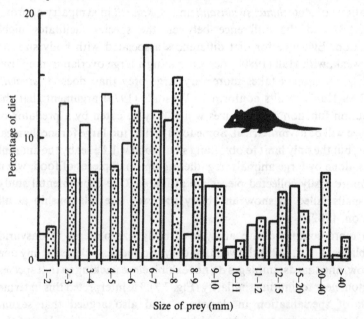

Charles-Dominique *et al.* (1981) argued that the sizes of members of the frugivorous didelphid community at Cabassou, French Guyana (Fig. 7.8) reflect selection that has reduced competition for fruits and animal prey of different sizes. Unfortunately, we are unable to comment further on this hypothesis as no quantitative data on differences between prey taken by

Table 7.7. *Ratio of sexual dimorphism (male/female) in the head–body length and basicranium length of* Antechinus *subspecies*

Subspecies	Locality	Dimorphism	
		Head–body length	Basicranium length
A. leo	Queensland	1.15	1.08
A. flavipes flavipes	Queensland	1.13	1.08
A. flavipes rubeculus	Queensland	1.13	1.05
A. stuartii adustus	Queensland	1.09	1.04
A. stuartii stuartii	Queensland	1.14	1.04
A. stuartii stuartii	New South Wales	1.08	1.04

Original data from Van Dyck (1980, 1982a).

Fig. 7.7. Body weight separation shown for dasyurid species assemblages found in different habitats in south-western Australia. Diamonds denote mean body mass. Numbers are ratios of body masses of species pairs. After Fox (1982b).

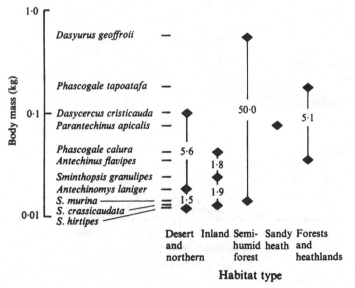

the frugivores are presented, and other differences in prey type, such as the phenological differences discussed in the section on mutualism, would confound any analysis of this problem.

Despite the frequency with which hypotheses of size-related limiting similarity are invoked, these hypotheses suffer from a number of severe problems. For example, Felsenstein (1981) argued that the number of species in an assemblage might be determined by genetic rather than ecological constraints, and called for a multidisciplinary framework for the examination of diversity. Far more important criticisms concern the uncritical acceptance of these postulations which do not have adequate statistical rigor or null hypothesis formulation (Roth, 1981; Simberloff, 1981; Simberloff & Boecklen, 1981; Wiens & Rotenberry, 1981). Certainly there does not appear to be a consistent size ratio of 1.3 and we may confidently reject Hypothesis 1 given above (Simberloff & Boecklen, 1981). Nor do the majority of statistical tests of constant size ratios such as those proposed by Roth (1979) and Barton & David (1956) support Hypothesis 2, even when applied to data originally interpreted as being supportive (Roth, 1981; Simberloff & Boecklen, 1981). This renders Hypothesis 3 unnecessary, and undermines the theoretical basis for the hypotheses as applied to marsupial communities.

Fig. 7.8. Distribution of head-body lengths for frugivorous mammals at Cabassou, French Guyana. (*a*) Arboreal species; (*b*) understorey species. After Charles-Dominique *et al.* (1981).

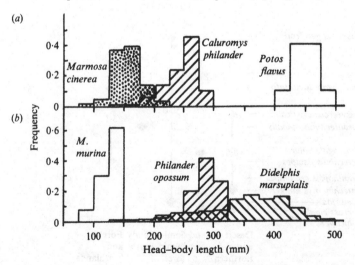

Null hypotheses in community ecology

The failure of empirical data cited in support of theory to satisfy statistical criteria is particularly damning given the tendency of scientists to publish positive but not negative results (Selven & Stuart, 1966). However, such inconsistency illustrates a problem of central importance in many of the controversies currently plaguing community ecology. During the optimistic explosion of mathematical theory in ecology, many field ecologists were happy to accept this theory without adequate empirical verification. For example, limiting similarity has been widely accepted in spite of empirical evidence to the contrary. Community data based on small samples collected over a short time period have also been used in support of theory, ignoring the effects of temporal variation (for a critique see Rotenberry & Wiens, 1980a; Wiens, 1981). A review of the attempts to introduce greater statistical rigor into analysis of community pattern would be premature at this stage; as this is such an active area any analysis is out-of-date as it is written. Suffice it to say that the empirical work and null hypothesis modelling of several workers has failed to support the widespread use of competition in explaining community patterns (e.g. Connor & McCoy, 1979; Connor & Simberloff, 1979; Strong, Szyska & Simberloff, 1979; Rotenberry & Wiens, 1980b; Strong, 1980; Wiens & Rotenberry, 1980; Bloom, 1981; Simberloff, 1981). This has prompted advocates of the importance of competition to attack the statistics used by their critics, or to introduce null models of their own (e.g. Grant & Abbott, 1980; Hendrickson, 1981; Diamond & Gilpin, 1982; Gilpin & Diamond, 1982; Wright & Biehl, 1982). In one instance, a critique of a critique is already available (Strong & Simberloff, 1981). While the vigor and intensity of this debate is exhausting to those trying to maintain it within some perspective, its results will inevitably be beneficial. Although proponents of 'no competition' and 'competition' models are at the moment divided into clear camps, hope of some reconciliation is apparent in recent contributions which suggest that both situations may pertain, and that the intensity and mode of coevolutionary patterning may vary from taxon to taxon (Lawton & Strong, 1981; Grossman, 1982).

Small mammal communities in south-eastern Australia

One attempt to utilise null models in the study of community relationships was Fox's (1981) analysis of the relationship between niche parameters and species richness in a community of five murid rodent species and two dasyurid marsupial species in an eastern Australian coastal heathland. Having derived measures of spatial niche breadth, overlap and separation (Fig. 7.9) on the basis of trapping success during the non-breeding season on a grid of trap sites, he compared the results with simulated random values. Three simulations were performed, with differing constraints:

(1) Within-patch simulation permitted testing of the hypothesis that field niche parameters do not differ significantly from those predicted for non-interacting species with random spatial distribution within patches which differed according to some floristic or structural criterion.

(2) Total-patch simulation permitted testing of the hypothesis that field niche parameters do not differ from those predicted for non-interacting species with random distribution over the entire plot, independent of patch locations.

(3) Equal simulation. In the preceding simulations the relative frequency of each species was held to the same value as occurred in the field data. To remove this constraint seven equally-abundant species were randomly distributed over the entire plot. This equal simulation tested the hypothesis that field niche parameters do not differ from those predicted for randomly-distributed, equally-abundant, non-interacting species.

These null hypotheses consequently represent a hierarchy of constraints which increasingly resemble the field data (Table 7.8). Fox (1981) hoped they would enable partitioning of the contributions from the various components examined. The various niche parameters measured in the field

Fig. 7.9. Resource utilisation for two species coexisting in a one-dimensional niche system showing niche breadth, overlap (stippled area) and separation. $\bar{x}$ denotes the mean; SD denotes one standard deviation. After Begon and Mortimer (1981).

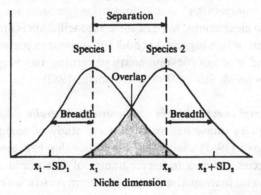

showed different levels of similarity to the field data. Mean spatial niche breadth simulations did not differ significantly from field values, implying that niche breadth is not affected by competition. Mean spatial niche separation simulations increasingly resembled the field data as constraints were added. However, for any value of species richness there were not significant differences in the separations, and the results of comparisons of the slopes for mean spatial separation were equivocal (Fox, 1981). The results for mean overlap were much more distinct. The slopes for mean

Table 7.8. *Constraints placed on simulations of small mammal distributions by Fox (1981). All simulations assume non-interaction. Note the hierarchical arrangement*

	Simulation		
	Within-patch	Total-Patch	Equal
Random distribution within patches	+	−	−
Capture frequency of each species as for field data	+	+	−

Fig. 7.10. Mean spatial niche overlap estimates for within-patch simulation (circles), total-patch simulation (squares) and actual field values (triangles) as a function of species richness of small mammals at Myall Lakes, New South Wales. The slope for field values differs significantly from each of the slopes for the simulations. After Fox (1981).

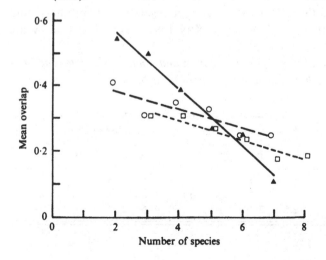

spatial niche overlap declined much more quickly with increasing species richness than in the simulations (Fig. 7.10).

This study illustrates a crucial point. Simulations which were based on non-interaction between species showed that niche breadth and mean overlap exhibit negative relationships with species richness while mean separation and total overlap show positive relationships. As Fox (1981) points out, such relationships have been used in the past to infer interspecific competition and in support of theory without the appropriate stochastic comparisons.

While Fox's (1981) analysis represents an important step towards increased rigor in the analysis of the effects of competition on community structure, it does not fill other criteria that are necessary for demonstration of the coevolutionary consequences of competition. First, no resource has been shown to be limiting for all the species involved. The animals differ greatly in the foods they consume (Braithwaite *et al.*, 1978), and the only alternative parameter considered by Fox (1981) which might be limiting is space. Yet a hypothesis of limited space is not supported by the observation that contiguous patches of identical vegetation differ by as much as three-fold in the density of animals they support (Table 7.9).

It is not particularly fruitful to expect all species to respond to the same niche dimension. Schoener (1974) argued that there is a correlation between the number of species in a community and the number of niche dimensions involved in the partitioning of resources. Thus species which are not separated along one niche dimension are separated along another. This is called niche dimension complementarity (Schoener, 1974). Other niche dimensions important to the small mammals include diet (Braithwaite *et al.*, 1978) and the post-fire succession of vegetation in both heathland and forest (Cockburn, 1978; Fox & McKay, 1981; Fox, 1982a). Braithwaite

Table 7.9. *Mean captures per trap station and number of species captured in contiguous patches of similar microhabitat, arbitrarily divided by Fox (1981) to satisfy statistical criteria*

Patch type	Mean no. of captures				Number of species			
	A	B	C	D	A	B	C	D
Short wet heath	1.0	2.2	—	—	2	3	—	—
Tall dry heath	—	2.7	2.8	1.0	—	5	5	4
Tree heath	2.5	3.2	—	—	6	7	—	—

et al. (1978) originally proposed that there were five food niches in heathland and forest communities:

(1) Generalist omnivore, including *Rattus fuscipes* and feral *R. rattus*.
(2) Above-ground generalist herbivore, including *Pseudomys* spp.
(3) Specialist herbivore, including *Rattus lutreolus* and *Mastacomys fuscus*.
(4) Soil-foraging insectivore, including *Antechinus swainsonii*, *A. minimus*, the Peramelidae, *Potorous tridactylus* and *Mus domesticus*.
(5) Small scansorial insectivore, including *A. stuartii*, *A. flavipes*, *Sminthopsis* spp. and *M. domesticus*.

Advances in our knowledge of the diets of these species (Chapter 2 and Cockburn, 1981a; Cockburn *et al.*, 1981) suggest that this view requires substantial modification. All the generalist species show seasonal variation in diet which leads to simultaneous exploitation of resources by species from a different niche according to the scheme proposed by Braithwaite *et al.* (1978) (e.g. *R. fuscipes*, *Pseudomys* spp., the Peramelidae and *P. tridactylus* all use hypogeous sporocarpic fungi heavily during winter). This again emphasises the importance of temporal variability in the analysis of competition (Wiens, 1981). A further form of temporal variability is succession over several years, which is of considerable importance in the fire-perturbed environments of south-eastern Australia. For example, Fox (1982a, c) presents evidence that *M. domesticus*, *Sminthopsis murina* and *Pseudomys* spp. are most abundant early in succession, while *A. stuartii* and *Rattus* spp. prefer mature habitats (Fig. 7.11).

'*The Ghost of Competition Past*'

The fact that species differ in diet and occupancy of seral stages need not imply competition. As Begon & Mortimer (1981) point out (p. 72), in the absence of direct experimental evidence:

> it is impossible to distinguish whether these data indicate inter-specific competition (first interpretation), its evolutionary avoidance (second interpretation), or its total non-existence now and in the past (third interpretation); and this would be true of almost all examples of apparent interspecific competition in the field. There are undoubtedly cases of current resource partitioning amongst competitors; and there are undoubtedly cases in which species ecologies have been moulded by interspecific competition in the past. But differences between species are *not*, in themselves,

indications of the ways in which those species coexist; and interspecific competition *cannot* be studied by the mere documentation of these interspecific differences.

Connell (1980) has lampooned the frequent inference of historical competition as an unsubstantiated invocation of the 'Ghost of Competition Past'. He argues that coevolutionary changes require prolonged co-occurrence, which will be more likely to have occurred when interacting species sought out each other (e.g. through mutualism, predation and parasitism), than when the co-occurrence is the consequence of sharing a common resource (competition). Co-occurrence will be reduced when the environment varies in space and time, which is relevant to the seasonal and successional effects described above. High species diversity is also associated with changing species composition, and Connell (1980) argues that coevolution should be more likely (i) in pairs of species of different trophic levels than in pairs competing on the same trophic level; and (ii) in communities with low species diversity in which there are low rates of change of species competition.

These problems led Connell (1980) to define a number of conditions necessary for justifying the inference of coevolution between competitors.

Fig. 7.11. The response of mammalian species to changes in vegetation regenerating after fire at Myall Lakes, New South Wales. (a) Wet habitat; (b) dry habitat. Question marks denote gap in sampling. After Fox (1982a).

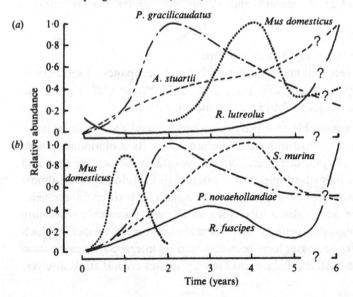

One must show (i) that divergence has actually occurred; (ii) that competition, rather than some other mechanism, is responsible; and (iii) that it has a genetic basis. Field experiments designed to demonstrate conditions (ii) and (iii) are described in Table 7.10, but these have only rarely been performed, and never for a community of higher vertebrates.

Summary

While competition might conceivably lead to coevolutionary divergence in the use of resources by competing species, it is not clear how frequently this may occur in natural communities. Several problems of technique have arisen in the attempt to unite empirical data and theory and these cloud the analysis of the importance of competition. They include:

(1) The absence of experimental evidence which demonstrates competition or the existence of some mutually limiting resource. (Dickman (1984) points out that all published evidence concerning competition between marsupials is inferential.)

(2) The assumption that differences in the use of resources implies competition and coevolution either now or in the past.

(3) Data dredging which leads to the reporting of results favourable to theory and not those unfavourable to theory.

(4) Failure to use null hypotheses, and appropriate statistical criteria for rejection of the null hypotheses.

(5) Failure to consider temporal variation, either seasonal or successional.

(6) Occasional refusal to abandon hypotheses in spite of negative evidence.

None of the marsupial studies we have cited are free of all these difficulties. Many of these problems are common to other areas of ecology and science generally, and just because recent criticism has focused on competition theory, workers in other areas should not be complacent about the assumptions they make in invoking evolutionary theory. One aspect of community ecology which makes it particularly intractable is the difficulty of applying evolutionary concepts applicable to individuals to the problem of interactions between species. Coevolution undoubtedly occurs, but it is responsible for only some of the properties of communities, and definition of the conditions which restrict its occurrence presents one of the major problems confronting theorists. Empirical evidence which aids this task will be invaluable, but must test rather than blindly accept mathematical theory.

Table 7.10. *Field experiments designed to demonstrate that competition is the source of coevolution (Proposition I), and that divergence attributable to competition has a genetic basis (Proposition II). If the answers to the questions posed under treatments 1, 2, 5 and 6 are affirmative, the propositions are acceptable (X_a, allopatric populations of species X; X_s, sympatric populations of species X; Y, the species presumed to be the competitor)*

	Species Y present naturally (sympatric locality)		Species Y absent naturally (allopatric locality)
	Species Y not removed	Species Y removed	
Proposition I is tested by observing changes in the breadth of the niche of X_a	Treatment 1 (X_a transplanted, X_s removed) Is X_a niche compressed so that it is significantly narrower than X_a niche in Treatment 2?	Treatment 2 (X_a transplanted, X_s removed) Does X_a niche remain broader than in Treatments 1 and 5?	Treatment 3 (X_a left in place as a control)
Proposition II is tested by observing changes in the breadth of the niche of X_s. (It need not change for Proposition II to be acceptable)	Treatment 4 (X_s left in place as a control)	Treatment 5 (X_s left in place) If X_s niche expands, is it always narrower than X_a niche in treatment 2?	Treatment 6 (X_s transplanted, X_a removed) If X_s niche expands, is it always narrower than X_a niche in treatment 3?

After Connell (1980).

Island biogeography and conservation

The number of species on an island or isolated patch tends to increase with area. The best statistical model of this relationship is a power function:

$$\log S = \log c + Z \log A$$

where S is the number of species, A is the area and c and Z are fitted constants (Z is a dimensionless parameter which typically varies between 0.18 and 0.35, while c depends on the taxonomic group used) (Diamond & May, 1981). Preston (1962a, b) and MacArthur & Wilson (1967) suggested that S is set by a dynamic balance between immigration and extinction. The immigration rate was proposed to decrease with increasing isolation and the extinction rate to increase with decreasing area, generating an equilibrium species number (Fig. 7.12). This hypothesis is still untested. Extrapolations of this theory have attempted to predict the characters of species which are either good immigrants or resistant to extinction, and whether island communities possess structure which reflects these relationships (Diamond, 1975; Williamson, 1981). These developments are among those which are currently subject to the controversy described in the section on competition.

The supposed relationships of species richness to area of habitat,

Fig. 7.12. Relative number of species on small distant islands (S_1) and large close islands (S_2) predicted by the MacArthur & Wilson (1967) equilibrium theory of island biogeography. After Ricklefs (1979).

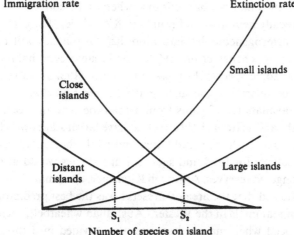

Number of species on island

extinction rate and degree of isolation have led a number of authors to argue that the theory of island biogeography can be applied to the design of nature reserves. Of course, this is only true if the objective of a nature reserve is to maximise species richness. The rules of design which have been most widely promoted are as follows:

(1) A large reserve will maintain more species than a small one.

(2) A single large reserve will maintain more species than two or more small ones with a total area equal to that of the single large one, because of isolation and area effects.

(3) Clumped fragments of habitat are better than widely dispersed fragments, because of isolation effects.

(4) Corridors or stepping-stones of natural habitat between reserves may be important for some species as routes for dispersal.

(5) A round design is best because an elongated reserve design may cause dispersal rates to outlying parts of the reserve from more central parts to be too low to overcome local extinctions.

(See also Fig. 7.13.)

Few conservationists would argue with rules (1) and (4). In a world where natural environments are being rapidly destroyed, the larger the reserve, the greater its conservation value, unless its economic costs hamper some other aspect of reserve management to the detriment of the whole. Dispersal corridors or stepping-stones are particularly valuable for migratory species (Margules, Higgs & Rofe, 1982).

By contrast, recent criticisms have been levelled at the utility of the remainder of these proposals (Margules *et al.*, 1982; Simberloff & Abele, 1982). Several small reserves have been shown to contain more species than a single large reserve on a number of occasions, whereas the converse has never been unequivocally demonstrated (Simberloff & Abele, 1982). This may be because in heterogeneous habitat, more habitat patches will be incorporated (see e.g. Kitchener *et al.*, 1980a), or because edge habitat facilitates survival of some species. For example, Smith (1982b) showed that a high population density of the sugar glider *Petaurus breviceps* was associated with the availability of gums from a roadside strip of *Acacia mearnsii* and of moths and scarabaeid beetles that have larval stages which develop in pastures adjacent to the study area and adult stages that are dependent upon eucalypts for food and shelter. Edge species would also be encouraged in elongated reserves. Although Kitchener, Chapman, Muir & Palmer (1980b) showed that the area of a reserve was the best predictor of species richness of mammals in the Western Australian wheatbelt, their analyses were enhanced when mammal species were divided into those favouring disturbed and those favouring undisturbed habitats, leading Kitchener (1982) to conclude that the autecological requirements of species need to be understood before prerequisites for nature reserves can be determined.

One situation for which large areas will be useful is the conservation of species which require a mosaic of stands of differing successional age. Cockburn (1978, 1979) and Fox (1982a) have shown that many heathland small mammals in south-eastern Australia prefer only certain stages of pyric succession (Fig. 7.11). Small areas are more likely to be of a single successional age, and are thus either too mature for these species or too young to provide refuges during a fire from which recolonisation can take place (Cockburn, 1978). This trend may tend to elevate species richness in the direction predicted by island biogeography theory, but again requires a knowledge of species idiosyncrasies for formulation of useful conservation strategies. The conservation programme would also be based on maximisation of habitat heterogeneity, whereas island biogeography theory assumes habitat homogeneity.

Extinction rates in other instances are poorly understood. In a comprehensive review, Gilbert (1980) concluded that there was no evidence that extinctions in reserves are due to decrease in area. Indeed, while the species–area relationship is supported by abundant evidence, the extinction–immigration equilibrium theory remains speculative, and a dubious basis for reserve design (Margules *et al.*, 1982).

Fig. 7.13. Geometric design strategies based on the equilibrium theory of island biogeography. After Diamond (1975). According to the theory, the design on the left is preferred as species number is maximised and extinction rate minimised. See text for definition of Strategies 1–5.

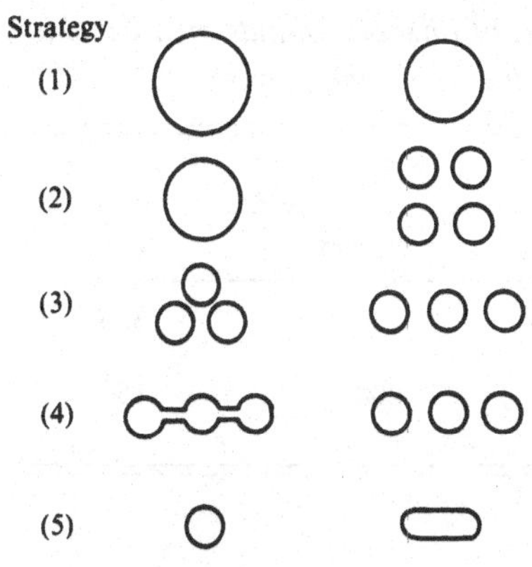

Simberloff & Abele (1982) point out that numerous other considerations minimise the utility of a general reductionist model for reserve design. These include:

(1) The many cases where the reserve attempts to achieve an aim other than maximisation of the number of conserved species.

(2) The fact that acquisition and subsequent management costs will always influence decisions on size and configuration.

(3) The fact that small isolated refuges might be insulated from catastrophe like epizootic disease.

(4) The consideration that if population size in a subrefuge was too small it might lead to inbreeding depression (Franklin, 1980; Shaffer, 1981).

(5) The naivety or arrogance of failure to consider the economic and social consequences of reserve design, particularly in the Third World, where the areas to be preserved support human populations dependent upon native vegetation for their existence.

Many of these issues are considered in the volume of papers edited by Soule & Wilcox (1980). They also lead us to agree with Margules *et al.* (1982) and Simberloff & Abele (1982) that island biogeography theory does not provide a general reductionist model which generates useful advice or predictions about reserve design, and that it may consequently provide little more than a warm feeling inside for those convinced that the incorporation of mathematical theory in some sense enhances the scientific validity of their work.

While the species–area relationship has been subject to occasional overinterpretation (Rey, Strong & McCoy, 1982), this does not mean that this relationship cannot occasionally be applied to specific questions of ecological management. Abbott (1980) showed that the distribution of

Table 7.11. *Median areas (in ha) of offshore islands in Australia, with respect to the presence of aborigines and macropods*

	Macropods	
	Present	Not present
Islands occupied or visited by aborigines	4800	300
Islands not occupied or visited by aborigines	800	150

After Abbott (1980).

macropods on Australian near-coastal islands with an area greater than 90 ha is non-random. Few islands adjacent to the north-western, northern and eastern coasts have macropods present, in contrast to islands near the mid- and south-western coasts and most of the southern coastline, and this is correlated with the lack of watercraft among aborigines living near the latter islands. Abbott (1980) argued that aborigines had exterminated macropods from islands they could visit, through direct predation, introduction of dingoes or habitat alteration. Macropods are present on 28% of islands with area 90–360 ha which are unused by aborigines in contrast to only 7% of islands of the same size which are used by aborigines.

The median area of islands which are used by aborigines and have at least one macropod species present is about six times larger than that of islands which have at least one macropod species present but are not visited or occupied by aborigines (Table 7.11). Abbott (1980) argued that the median size of islands used by aborigines and supporting a macropod population provides a clue to the minimum area necessary to support a population of macropods throughout the 9000 years since the isolation of the islands. This figure (about 4800 ha) could be used as a basis for the design of mainland reserves which are subject to aboriginal hunting and firing.

Summary

There are instances where evolutionary and ecological theory may be usefully applied to the design of conservation strategies, and recent marsupial studies illustrate some of these. However, a reductionist model derived from the equilibrium theory of island biogeography does not offer a general conservation strategy. Effective conservation will be influenced by a number of factors, of which pragmatic considerations and autecological information seem likely to remain most important.

8

Future directions

From the foregoing analyses it should be obvious that there are still major deficiencies in our data base for marsupials. Although we would welcome any information, either published or unpublished, which helps to overcome these gaps, we believe the available data point to a number of important theoretical areas worthy of future investigation.

It is clear from Chapter 2 that it is important to distinguish between anecdotal observations on diet and those studies which carefully measure food intake, particularly in relation to food availability. Only a few studies have unequivocally demonstrated food preference by marsupials. In order to interpret the importance of particular food resources it will be necessary to attempt experimental manipulations of resources to determine which of them are limiting to particular populations.

In Chapter 3 we stress the need for increased recognition of allometric and developmental constraints on life history traits in marsupials. We also feel that it is important to consider traits of species in a holistic perspective rather than in terms of exigencies operating on individual species. Our data also point to the need for increasing synthesis of ecological, developmental and physiological approaches to marsupial biology. Specific areas worthy of further research identified in this chapter include the possibility of adaptive brood reduction in polytocous marsupials, measurement of the cost of lactation in eutherians versus marsupials, examination of the sources of the rapid pace of life found in the Peramelidae and interpretation of marsupial growth and development in terms of heterochrony.

In Chapter 4 we review the biology of three well-studied taxa of marsupials which feed predominantly on animal tissue. In the Dasyuridae it will be important to quantify the relationship between temporal changes in food availability and reproductive effort. Specific projects worthy of

222

further investigation include measurement of the levels of male reproductive investment and success in dasyurids which do and do not exhibit male die-off, and experimental analysis of the effect of early loss of a litter on those facultatively monoestrous species which may return to oestrus if their litter is lost. Establishment of generalities about the Didelphidae will be impossible until the latitudinal and altitudinal amplitude of population studies have been expanded to embrace populations outside the tropics. We have already alluded to the rapid pace of life in the Peramelidae. This family may also be especially suitable for studies of brood reduction. Further areas requiring investigation include measurement of dispersal in bandicoots in relation to habitat colonisation and the factors which control juvenile recruitment.

In Chapter 5 we examine the life histories and social behaviour of the herbivorous marsupials. Future studies should attempt a theoretical integration of the interaction between life history and social behaviour in possums, particularly with respect to the frequency of occurrence of monogamy in these species. It will be valuable to assess the roles and kinship of subordinate and reproductive dominant males which associate in both *Gymnobelideus leadbeateri* and *Trichosurus vulpecula*. More analysis of the various advantages of embryonic diapause in different environments is also warranted. In addition, diapause has been frequently exploited by reproductive physiologists to induce reproductive synchrony and to time precisely the onset and termination of reproduction. This technique has not been adequately exploited by ecologists.

In Chapter 6 we illustrate the utility of the evolutionary approach to marsupial biology by illustrating some of our own work on *Antechinus*. Areas we are currently studying include sex-biased dispersal, allocation of resources by parents to the different sexes, parent–offspring conflict and evolution at the boundary between iteroparous and semelparous life histories.

Finally, in Chapter 7 we examine the effect of coevolution on the structure of marsupial communities, and are critical of the lack of quantification of proposed coevolutionary relationships. Particular attention should be paid to quantifying benefits accruing to plants from proposed plant/pollinator, plant/frugivore and plant/fungivore mutualism and increased scientific rigor and experimentation are desperately needed for analysis of competition between marsupial species.

There is obviously plenty of work to keep us busy for many years.

Appendix 1. Body dimensions of marsupials. Systematic arrangement follows Kirsch & Calaby (1977)

Marsupial	Body mass (g) ♂♂	Body mass (g) ♀♀	Head–body length (mm) ♂♂	Head–body length (mm) ♀♀	Tail length (mm) ♂♂	Tail length (mm) ♀♀	Source
Family Didelphidae							
Marmosa (Marmosa) alstoni			179		238		Hall & Kelson (1959)
canescens			122		151		Hall & Kelson (1959)
cinerea			144		207		Tate (1933)
constantiae			160		210		Tate (1933)
cracens			103		132		Tate (1933)
domina			166		250		Tate (1933)
emiliae			75		142		Tate (1933)
fuscata			134		151		Tate (1933)
germana			187		235		Tate (1933)
handleyi			104		129		Tate (1933)
impavida			126		162		Hall & Kelson (1959)
incana			130		170		Tate (1933)
invicta			104		136		Hall & Kelson (1959)
juninensi			110		133		Tate (1933)
lepida			99		140		Thomas (1888)
leucastra			126		145		Tate (1933)
mapiriensis			162		202		Tate (1933)
mexicana			168		170		Hall & Kelson (1959)
murina		57	144		188		Hall & Kelson (1959); Grand (1983)
noctivaga			133		175		Tate (1933)
ocellata			145		175		Tate (1933)
parvidens			105		130		Tate (1933)
phaea			132		179		Tate (1933)
quichua			166		142		Tate (1933)
rapposa			160		210		Tate (1933)
regina			229		226		Tate (1933)
robinsoni	47	122	153		185		Hall & Kelson (1959); McNab (1978b)
rubra			150		195		Tate (1933)
scapulata			160		190		Tate (1933)

Species				Reference
tylerina	151		171	Tate (1933)
xerophila	131		163	Tate (1933)
yungasensis	136		169	Tate (1933)
M. (Stegomarmosa) andersoni	—		—	
M. (Thylamys) aceramarcae	83		112	Tate (1933)
agilis	110		137	Tate (1933)
agricolai	—		—	
dryas	97		147	Tate (1933)
elegans	95		—	Thomas (1888)
formosa	68		55	Tate (1933)
grisea	129		—	Thomas (1897)
karimii	—		—	
marica	102		131	Tate (1933)
microtarsus	75	13	131	Tate (1933); McNab (1978b)
pusilla	92		—	Thomas (1899)
tatei	—		—	
unduaviensis	87		115	Tate (1933)
velutina	86		—	Thomas (1897)
Monodelphis adusta	108		60	Hall & Kelson (1959)
americana	107	73	—	Thomas (1888)
brevicaudata	130		—	Thomas (1888); Grand (1983)
dimidiata	143		—	Thomas (1888)
domestica	131		—	Thomas (1888)
handleyi	—		—	Thomas (1888)
henseli	106		—	Thomas (1888)
iheringa	77		43	Thomas (1888)
maraxina	134		79	Tate (1933)
orinoci	—		—	Thomas (1888)
scalops	133		—	Thomas (1888)
sorex	67		—	Thomas (1888)
touan	—		—	Thomas (1888)
unistriata	140		—	Thomas (1888)
Lestodelphys halli	138		—	Marshall (1977)
Metachirus nudicaudatus	265	391	332	Hall & Kelson (1959); Grand (1983)
Didelphis albiventris	—		—	
marsupialis	471	1165	477	Hall & Kelson (1959); McNab (1978b)
virginiana	497		415	Hall & Kelson (1959); McNab (1978b)
Philander mcillhenyi	—		—	
opossum	259	381 / 341	268	Hall & Kelson (1959); Grand (1983)

Appendix 1 (cont.)

Marsupial	Body mass (g) ♂♂	Body mass (g) ♀♀	Head–body length (mm) ♂♂	Head–body length (mm) ♀♀	Tail length (mm) ♂♂	Tail length (mm) ♀♀	Source
Lutreolina crassicaudata	812		325		—		Marshall (1977); McNab (1982)
Chironectes minimus	946		285		383		Hall & Kelson (1959); McNab (1982)
Caluromys derbianus	329		231		442		Hall & Kelson (1959); McNab (1978b, 1982)
lanatus			270		—		Thomas (1888)
philander	340	285	290		—		Thomas (1888); Grand (1983)
Caluromysiops irrupta			217		—		Sanborn (1951)
Glironia criniger			—		—		
venusta			183				Marshall (1978b)
Family Microbiotheriidae							
Dromiciops australis			112		107		Marshall (1978a)
Family Caenolestidae							
Caenolestes caniventer			129		127		Anthony (1921)
canvelatus			124		132		Anthony (1923)
fuliginosus			123		110		Anthony (1923)
obscurus		40	111		111		Anthony (1923); McNab (1978b)
tatei			96		117		Anthony (1923)
Lestoros inca			90		120		Walker (1968)
Rhyncholestes raphanurus			100		76		Walker (1968)
Family Dasyuridae							
Antechinus bellus	55	34	134	117	114	100	Strahan (1983)
flavipes	56	34	121	105	100	86	Strahan (1983)
godmani	95	55	143	122	126	104	Strahan (1983)
leo	70	36	139	121	120	100	Strahan (1983)
melanurus			140		—		Ziegler (1982)
minimus	65	42	120	110	80	75	Strahan (1983)
naso			145		—		Ziegler (1982)

Species							Reference
stuartii	35	20	100	92	95	85	Strahan (1983)
swainsonii	65	41	128	116	107	92	Strahan (1983)
wilhelmina			130				Ziegler (1982)
Dasycercus cristicauda	75–170	60–95	125–220	125–170	75–125	75–100	Strahan (1983)
Dasykaluta rosamondae	25–40	20–30	95–110	90–110	55–70	60–65	Strahan (1983)
Dasyuroides byrnei	120	100	165	150	120	115	Strahan (1983)
Dasyurus geoffroii	1300	850	347	305	251	250	Strahan (1983)
masculatus	<7000	<4000	380–759	350–450	370–550	340–420	Strahan (1983)
viverrinus	1300	880	370	340	240	220	Strahan (1983)
Murexia longicaudata			285	200			Ziegler (1982)
rothschildi			170				Ziegler (1982)
Myoictis melas			250				Ziegler (1982)
Neophascogale lorentzii			—		—		
Ningaui ridei	6.5–13.0		57–71	140	59–71	95	Strahan (1983)
timealeyi	2.0–9.4		46–57		59–79		Strahan (1983)
Parantechinus apicalis	60–100	40–75	145	57–98	105–115		Strahan (1983)
bilarni	12–44	12–34	60–100	101	82–115	132	Strahan (1983)
Phascogale calura	60	43	113	179	141	182	Strahan (1983)
tapoatafa	199	145	202		196		Strahan (1983)
Phascolosorex doriae			230				Ziegler (1982)
dorsalis			175				Ziegler (1982)
Planigale gilesi	11.5	6.9	76	68	64	60	Strahan (1983)
ingrami	4.2	4.3	59	59	59	52	Strahan (1983)
maculatus	12	10	81	79	79	76	Strahan (1983)
novaegineae	—	—	96	78	80	67	Ziegler (1982)
tenuirostris	6.8	5.3	65	62	56	55	Strahan (1983)
Pseudantechinus macdonnellensis	25–45	20–40	95–105	95–105	75–80	75–85	Strahan (1983)
Satanellus albopunctatus			360				Ziegler (1982)
hallucatus	400–900	300–500	123–310	125–300	127–308	200–300	Strahan (1983)
Sarcophilus harrisii	8000	6000	652	570	258	244	Strahan (1983)
Sminthopsis (Sminthopsis) butleri			88		90		Strahan (1983)
crassicaudata	15		75		55		Strahan (1983)
granulipes	25		83		60		Strahan (1983)
hirtipes		18	78	74	86	90	McKenzie & Archer (1982)
leucopus	26	20	100	92	83	73	Strahan (1983)
longicaudata	15–20	15–20	80–100	80–90	200–210	180–190	Strahan (1983)
macroura	20		85		95		Strahan (1983)
murina	16–28	16–22	76–104	64–92	70–99	68–92	Strahan (1983)
ooldea	11	11	72	72	79	76	Strahan (1983)

Appendix 1 (cont.)

Marsupial	Body mass (g) ♂♂	Body mass (g) ♀♀	Head–body length (mm) ♂♂	Head–body length (mm) ♀♀	Tail length (mm) ♂♂	Tail length (mm) ♀♀	Source
psammophila	10	—	91–114	—	114–129	—	Strahan (1983)
douglasi	—	40	98	—	—	104	Strahan (1983)
virginiae	—	10	68	69	66	65	McKenzie & Archer (1982)
youngsoni			85	85	130	120	Strahan (1983)
S. (Antechinomys) laniger	c. 30	c. 20					Strahan (1983)
Family Myrmecobiidae							
Myrmecobius fasciatus	495	416	243	247	181	175	Strahan (1983)
Family Thylacinidae							
Thylacinus cynocephalus			1000–1300		500–650		Strahan (1983)
Family Notoryctidae							
Notoryctes typhlops	40–70		121–159		21–26		Strahan (1983)
Family Peramelidae							
Chaeropus eucaudatus			230–260		100–150		Strahan (1983)
Echymipera clara			450		—		Ziegler (1982)
kaluba			385		—		Ziegler (1982)
rufescens	500–2000	500–1400	300–450		75–100		Strahan (1983)
Isoodon auratus	267–670		215		100		Strahan (1983)
macrourus	2100	1100	400	350	170	130	Strahan (1983)
obesulus	850	700	330	300	120	110	Strahan (1983)
Microperoryctes murina			150–175		105–110		Ziegler (1977)
Perameles bougainville	220		240	275	90	135	Ziegler (1977)
eremiana			235		118		Strahan (1983)
gunnii	650		305		87		Strahan (1983)
nasuta	850–1100		310–425		120–155		Strahan (1983)
Peroryctes longicauda	< 4700		305		—		Ziegler (1982)
papuensis			200		—		Ziegler (1982)
raffrayanus			370		—		Ziegler (1982)
broadbenti	—	—	—		—		
Rhynchomeles prattorum			—		—		Ziegler (1977, 1982)

Taxon							Reference
Family Thylacomyidae							
Macrotis lagotis	1000–2500	800–1100	300–550	290–390	200–290	200–278	Strahan (1983)
leucura	360–435	311	240–270	200–240	125–170	120–150	Strahan (1983)
Family Phalangeridae							
Phalanger carmelitae			475				Ziegler (1982)
celebensis			—		—		Ziegler (1982)
gymnotis			515		—		Ziegler (1982)
interposites			470		—		Ziegler (1982); Strahan (1983)
maculatus	1500–3600		348	650	315–430		Ziegler (1982); Strahan (1983)
lullulae			410				Ziegler (1982)
orientalis	1475–2200		350–470		280–350		Ziegler (1982); Strahan (1983)
rufoniger					—		
ursinus					—		
vestitus			480				Ziegler (1982)
Trichosurus arnhemensis	1600	1300	420	400	260	270	Strahan (1983)
caninus	2500–4500		400	500	340–420		Strahan (1983)
vulpecula	2000–4500	1500–3500	350	550	250–400		Strahan (1983)
Wyulda squamicaudata			c. 400		c. 300		Strahan (1983)
Family Tarsipedidae							
Tarsipes rostratus	9	11	68	60–80	83	60–80	Strahan (1983)
Family Burramyidae							
Acrobates pygmaeus	10.5–16.5		65–80	60–80	70–80		H. Frey (personal communication)
Burramys parvus	43.5	40	115	110	148	140	Strahan (1983)
Cercartetus caudatus	30		106		135		Strahan (1983)
concinnus	13		81		86		Strahan (1983)
lepidus	7		64		71		Strahan (1983)
nanus	24		91		89		Strahan (1983)
Distoechurus pennatus			125		—		Ziegler (1982)
Family Petauridae							
Dactylopsila (Dactylonax) palpator			255		—		Ziegler (1982)
D. (Dactylopsila) megalura			240		—		Ziegler (1982)
tatei			220		—		Ziegler (1982)
trivirgata	315		263		325		Strahan (1983)
Gymnobelideus leadbeateri	122–133		160		172		Strahan (1983)
Petauroides volans	900–1700		350–450		450–600		Strahan (1983)
Petaurus abidi			265				Ziegler (1981)
australis	450–700		280		433		Strahan (1983)
breviceps	140	115	170	170	190	190	Strahan (1983)

Appendix 1 (cont.)

Marsupial	Body mass (g)		Head–body length (mm)		Tail length (mm)		Source
	♂♂	♀♀	♂♂	♀♀	♂♂	♀♀	
norfolcencis	230		210		270		Strahan (1983)
Pseudocheirus (Pseudocheirus)			285				Ziegler (1982)
canescens			320				
caroli			340				Ziegler (1982)
forbesi			349				Ziegler (1982)
herbertensis	1070		—		360		Strahan (1983)
mayeri							
peregrinus	700–1100		300–350		300–350		Strahan (1983)
schlegeli			230				Ziegler (1982)
P. (Pseudochirops) albertisii			340				Ziegler (1982)
archeri	1190		364		321		Strahan (1983)
corinnae			370		—		Ziegler (1982)
cupreus			460		—		Ziegler (1982)
dahli	1280–2000		334–375	349–383	200–220	207–266	Strahan (1983)
P. (Hemibelideus) lemuroides	966		330		355		Strahan (1983)
Family Macropodidae							
Subfamily Potoroinae							
Aepyprymnus rufescens	3000	3 500	385		360		Strahan (1983)
Bettongia gaimardi	1660		323		326		Strahan (1983)
lesueur			280–400[a]		215–300[a]		Strahan (1983)
penicillata	1300		330		310		Strahan (1983)
Caloprymnus campestris	797	981	268	267	314	335	Strahan (1983)
Hypsiprymnodon moschatus	529	511	230	233	145	140	Strahan (1983)
Potorous platyops			243		183		Strahan (1983)
tridactylus	1180	1020	380	340	235	228	Strahan (1983)
longipes	2100	1700	400		320		Strahan (1983)
Subfamily Macropodinae							
Dendrolagus bennettianus	13000		650	500–610	940	631–685	Strahan (1983)
dorianus			780		—		Ziegler (1982)
goodfellowi			770		—		Ziegler (1982)

							Reference
inustus	7400		555	480	668	700	Ziegler (1982)
lumholtzi	5900		790				Strahan (1983)
matschiei			660				Ziegler (1982)
ursinus			730				Ziegler (1982)
Dorcopsis atrata			620				Ziegler (1982)
hageni			690				Ziegler (1982)
veterum			680				Ziegler (1982)
Dorcopsulus macleayi			515				Ziegler (1982)
vanheurni			450				Ziegler (1982)
Lagorchestes asomatus			—		—		
conspicillatus	1600–4500		400–470	375	370–490	275	Strahan (1983)
hirsutus	1580	1740	330		270		Strahan (1983)
leporides	1300–2500	1300–3000	450		320		Strahan (1983)
Lagostrophus fasciatus			430		370		Strahan (1983)
Macropus agilis	19000	11000	800	650	770	640	Strahan (1983)
antilopinus	37000	17500	1064	805	815	692	Strahan (1983)
bernardus	21000	13000	683	646	609	575	Strahan (1983)
dorsalis	16000	6500	1525	1150	765	595	Strahan (1983)
eugenii	7500	5500	643	586	411	379	Strahan (1983)
fuliginosus	3000–53500	4500–27500	525–1225	528–931	425–1000	443–815	Strahan (1983)
giganteus	4000–66000	3500–32000	542–1212	512–1015	430–1090	446–842	Strahan (1983)
greyi			810	840	730	710	Strahan (1983)
irma	8000		1200		720		Strahan (1983)
parma	4100–5900	3200–4800	482–528	447–527	489–544	405–507	Strahan (1983)
parryi	16000	11000	<924	<755	941	781	Strahan (1983)
robustus	7250–46500	6250–25000	587–1085	573–831	551–901	534–749	Strahan (1983)
rufogriseus	18600–19700[a]	13800–14000[a]	782–823[a]	706–772[a]	768–797[a]	695–720[a]	Strahan (1983)
rufus	66000	26500	1150	1000	880	820	Strahan (1983)
Onychogalea fraenata	5000–6000	4000–5000	510–700	430–540	380–540	360–440	Strahan (1983)
lunata			371–508		153–330		Strahan (1983)
unguifera	7500	5800	600	570	660	630	Strahan (1983)
Peradorcas concinna	1300–1400		319		297		Strahan (1983)
Petrogale brachyotis	4500		480		490		Strahan (1983)
burbidgei	1258		322		276		Strahan (1983)
godmani	5000		495		551		Strahan (1983)
inornata	3700–4300[a]		470–509[a]		465–533[a]		Strahan (1983)
lateralis	4250–5700[a]		480–525		540–560[a]		Strahan (1983)
penicillata	5800–7500[a]		500–540[a]		540–610[a]		Strahan (1983)
persephone	5000–8000		520–640		600–680		Strahan (1983)

Appendix 1 (cont.)

Marsupial	Body mass (g) ♂♂	♀♀	Head–body length (mm) ♂♂	♀♀	Tail length (mm) ♂♂	♀♀	Source
rothschildi	5250		525		590		Strahan (1983)
xanthopus	6000–7000		580–600		520–690		Strahan (1983)
Setonix brachyurus	3600	2900	487	468	289	265	Strahan (1983)
Thylogale billardierii	7000	3900	630	560	417	320	Strahan (1983)
brunii			725		—		Strahan (1983)
stigmatica	5100	4200	492	463	433	357	Strahan (1983)
thetis	7000	3800	520	420	430	350	Strahan (1983)
Wallabia bicolor	17000	13000	756	697	761	692	Strahan (1983)
Family Phascolarctidae							
Phascolarctos cinereus	11800	7900	782	716	—	—	Strahan (1983)
Family Vombatidae							
Lasiorhinus latifrons	19000–32000		772–934		25–60		Strahan (1983)
krefftii			c. 1000		c. 50		Strahan (1983)
Vombatus ursinus	26000		985		c. 25		Strahan (1983)

ᵃ Range of means for subspecies.

Appendix 2. *Regression statistics for Chapter 3*

Table A2.1. *Regression statistics for allometric equations relating life history traits to maternal body mass*

Trait	Equation	n	R^2	P
$PI_{Marsupial}$	3.8	37	0.727	< 0.001
$PI_{Eutherian}$	3.9	31	0.566	< 0.001
$t_{cw (Marsupial)}$	3.12	22	0.173	< 0.050
$t_{cw (Eutherian)}$	3.10	23	0.466	< 0.001
$\alpha_{Marsupial}$	3.14	11	0.853	< 0.001
$\alpha_{Eutherian}$	3.18	6	0.850	< 0.010

Table A2.2. *Tests for differences between slope and elevation of allometric equations for marsupials and eutherians*

Trait	t_{slope}	df	P	$t_{elevation}$	df	P
PI	0.40	64	0.350	1.51	65	0.067
t_{cw}	1.52	41	0.068	15.06	42	< 0.001
α	2.14	13	0.026	not required		

Procedures follow Zar (1974).

References

Abbott, I. (1980). Aboriginal man as an exterminator of wallaby and kangaroo populations on islands round Australia. *Oecologia (Berlin)*, **44**, 347–54.

Adrian, J. (1976). Gums and hydrocolloids in nutrition. *World Review of Nutrition and Dietetics*, **25**, 186–216.

Alberch, P. (1980). Ontogenesis and morphological diversification. *American Zoologist*, **20**, 653–67.

Alberch, P. (1982). Developmental constraints in evolutionary processes. In *Evolution and Development*, ed. J. T. Bonner, pp. 313–32. Berlin, Heidelberg & New York: Springer-Verlag.

Alberch, P., Gould, S. J., Oster, G. F. & Wake, D. B. (1979). Size and shape in ontogeny and phylogeny. *Paleobiology*, **5**, 296–317.

Anderson, P. K. (1980). Evolutionary implications of microtine behavioral systems on the ecological stage. *Biologist*, **62**, 70–88.

Andrew, D. L. & Settle, G. A. (1982). Observations on the behaviour of species of *Planigale* (Dasyuridae, Marsupialia) with particular reference to the narrow-footed planigale (*Planigale tenuirostris*). In *Carnivorous Marsupials*, ed. M. Archer, pp. 311–24. Sydney: Royal Zoological Society of New South Wales.

Andrewartha, H. G. & Barker, S. (1969). Introduction to a study of the ecology of the Kangaroo Island wallaby, *Protemnodon eugenii* (Desmarest) within Flinders Chase, Kangaroo Island, South Australia. *Transactions of the Royal Society of South Australia*, **93**, 127–32.

Anthony, H. E. (1921). Preliminary report on Ecuadorean mammals. 1. *American Museum Novitates*, **20**, 1–6.

Anthony, H. E. (1923). Preliminary report on Ecuadorean mammals. 3. *American Museum Novitates*, **55**, 1–14.

Archer, M. (1974). Some aspects of reproductive behaviour and the male erectile organs of *Dasyurus geoffroii* and *D. hallucatus* (Dasyuridae: Marsupialia). *Memoirs of the Queensland Museum*, 17, 63–7.

Archer, M. (1975). *Ningaui*, a new genus of tiny dasyurids (Marsupialia) and two new species, *N. timealeyi* and *N. ridei*, from arid Western Australia. *Memoirs of the Queensland Museum*, 7, 237–49.

Archer, M. (1976). Revision of the marsupial genus *Planigale* Troughton (Dasyuridae). *Memoirs of the Queensland Museum*, 17, 341–65.

Archer, M. (1981a). Results of the Archbold expeditions. No. 104. Systematic revision of the marsupial dasyurid genus *Sminthopsis* Thomas. *Bulletin of the American Museum of Natural History*, **168**, 61–224.

Archer, M. (1981b). A review of the origins and radiations of Australian mammals. In *Ecological Biogeography of Australia*, Vol. III, ed. A. Keast, pp. 1437–88. The Hague: Dr. W. Junk.

Archer, M. (1982). Genesis: and in the beginning there was an incredible carnivorous mother. In *Carnivorous Marsupials*, ed. M. Archer, pp. vii–x. Sydney: Royal Zoological Society of New South Wales.

Archer, M. & Kirsch, J. A. W. (1977). The case for the Thylacomyidae and Myrmecobiidae. Gill, 1872, or why are marsupial families so extended? *Proceedings of the Linnaean Society of New South Wales*. **102**, 18–25.

Armstrong, J. A. (1979). Biotic pollination mechanisms in the Australian flora: a review. *New Zealand Journal of Botany*, **17**, 467–508.

Arnold, J. (1976). Growth and development of the chuditch, *Dasyurus geoffroii*. Ph.D. thesis. University of Western Australia.

Arnold, J. E. & Shield, J. (1970a). Growth and development of the chuditch (*Dasyurus geoffroii*). *Bulletin of the Australian Mammal Society*, **2**, 198.

Arnold, J. E. & Shield, J. (1970b). Oxygen consumption and body temperature of the chuditch (*Dasyurus geoffroii*). *Journal of Zoology, London*, **160**, 391–404.

Arthur, W. (1982). A developmental approach to the problem of variation in evolutionary rates. *Biological Journal of the Linnaean Society*, **18**, 243–61.

Asdell, S. A. (1964). *Patterns of Mammalian Reproduction*, 2nd edn. London: Constable.

Ashmole, N. P. (1963). The regulation of numbers of tropical oceanic birds. *Ibis*, **103b**, 458–73.

Aslin, H. (1974). The behaviour of *Dasyuroides byrnei* in captivity. *Zeitschrift für Tierpsychologie*, **35**, 187–208.

Atherton, R. G. & Haffenden, A. T. (1982). Observations on the reproduction and growth of the long-tailed pygmy possum, *Cercartetus caudatus* (Marsupialia: Burramyidae), in captivity. *Australian Mammalogy*, **5**, 253–9.

Atramentowicz, M. (1982). Influence du milieu sur l'activite locomotrice et la reproduction de *Caluromys philander* (L.). *Revue d'Ecologie (La Terre et la Vie)*, **36**, 373–95.

Attwill, P. M. (1980). Nutrient cycling in a *Eucalyptus obliqua* (L'Herit.) forest. IV. Nutrient uptake and nutrient cycling. *Australian Journal of Botany*, **28**, 199–222.

Austin, M. P. (1971). Role of regression analysis in plant ecology. *Proceedings of the Ecological Society of Australia*, **6**, 63–75.

Bailey, P. T. (1971). The red kangaroo, *Megaleia rufa* (Desmarest), in north-western New South Wales. I. Movements. *CSIRO Wildlife Research*, **16**, 11–28.

Bailey, P. T., Martensz, P. N. & Barker, R. (1971). The red kangaroo, *Megaleia rufa* (Desmarest), in north-western New South Wales. II. Food. *CSIRO Wildlife Research*, **16**, 29–39.

Baird, D. T., Horton, R., Longcope, C. & Tait, J. F. (1969). Steroid dynamics under steady-state conditions. *Recent Progress in Hormone Research*, **25**, 611–64.

Baker, H. G. (1960). The adaptation of flowering plants to nocturnal and crepuscular pollinators. *Quarterly Review of Biology*, **36**, 64–73.

Baker, H. G. (1976). "Mistake" pollination as a reproductive system with particular reference to the Caricaceae. In *Tropical Trees as Living Systems*, ed. P. B. Tomlinson & M. Zimmerman, pp. 57–82. Cambridge: Cambridge University Press.

Baker, H. G. & Baker, I. (1975). Studies of nectar constitution and pollinator–plant coevolution. In *Coevolution of Animals and Plants*, ed. L. E. Gilbert & P. H. Raven, pp. 100–40. Austin: University of Texas Press.

Bamford, J. M. (1973). Population statistics and their relation to the control of opossums in indigenous forest. In *Assessment and Management of Introduced Animals in New Zealand Forests*. New Zealand Forest Service, Forest Research Institute Symposium, **14**, 38–43.

Barker, I. K., Beveridge, I., Bradley, A. J. & Lee, A. K. (1978). Observations on spontaneous stress-related mortality among males of the dasyurid marsupial *Antechinus stuartii* Macleay. *Australian Journal of Zoology*, **26**, 435–44.

Barker, S. (1974). Studies on seasonal anemia in the Rottnest Island quokka, *Setonix brachyurus* (Quoy and Gaimard) (Marsupialia: Macropodidae). *Transactions of the Royal Society of South Australia*, **98**, 43–8.

Barnett, J. L., How, R. A. & Humphreys, W. F. (1982). Habitat effects on organ weights, longevity and reproduction in the mountain brushtail possum, *Trichosurus caninus* (Ogilby). *Australian Journal of Zoology*, **30**, 23–32.

Barton, D. E. & David, F. N. (1956). Some notes on random ordered intervals. *Journal of the Royal Statistical Society, B*, **18**, 79–94.

Baudinette, R. V. (1980). Physiological responses to locomotion in marsupials. In *Comparative Physiology: Primitive Mammals*, ed. K. Schmidt-Nielsen, L. Bolis & C. R. Taylor, pp. 200–12. Cambridge: Cambridge University Press.

Baudinette, R. V. (1982). The energetics of locomotion in dasyurid marsupials. In *Carnivorous Marsupials*, ed. M. Archer, pp. 261–5. Sydney: Royal Zoological Society of New South Wales.

Beattie, A. J. & Culver, D. C. (1982). Inhumation: how ants and other invertebrates help seeds. *Nature, London*, **297**, 627.

Begg, R. J. (1981a). The small mammals at Little Nourlangie Rock, N.T. II. Ecology of *Antechinus bilarni*, the sandstone antechinus (Marsupialia: Dasyuridae). *Australian Wildlife Research*, **8**, 57–72.

Begg, R. J. (1981b). The small mammals at Little Nourlangie Rock, N.T. III. Ecology of *Dasyurus hallucatus*, the northern quoll (Marsupialia: Dasyuridae). *Australian Wildlife Research*, **8**, 73–85.

Begon, M. & Mortimer, M. (1981). *Population Ecology: A Unified Study of Animals and Plants*. Oxford: Blackwell Scientific Publications.

Bell, B. D. (1981). Breeding and condition of possums, *Trichosurus vulpecula*, in the Orongorongo Valley, near Wellington, New Zealand, 1966–1975. *Zoological Publications from Victoria University of Wellington*, **74**, 87–139.

Bell, H. M. (1973). The ecology of three macropod marsupial species in an area of open forest and savannah woodland in north Queensland, Australia. *Mammalia*, **37**, 527–44.

Berg, R. Y. (1975). Myrmecochorous plants in Australia. Their dispersal by ants. *Australian Journal of Botany*, **23**, 475–508.

Berger, P. J. (1966). Eleven-month "embryonic diapause" in a marsupial. *Nature, London*, **211**, 435–6.

Bergeron, J. (1980). Importance des plantes toxiques dans le regime alimentaire de *Microtus pennsylvanicus* a deux etapes opposees de leur cycle. *Canadian Journal of Zoology*, **58**, 230–8.

Besançon, F. (1983). Litter size and foetal mortality in *Crocidura russula* (Insectivora: Soricidae) and comparison with other Crocidurinae. *Acta Zoologica Fennica*, in press.

Blackall, S. (1980). Diet of the Eastern Native-Cat, *Dasyurus viverrinus* (Shaw), in southern Tasmania. *Australian Wildlife Research*, **7**, 191–7.

Bloom, S. A. (1981). Specialization and non-competitive resource partitioning among sponge-eating dorid nudibranchs. *Oecologia (Berlin)*, **49**, 305–15.

Blueweiss, L., Fox, H., Kudzma, V., Nakashima, D., Peters, R. & Sam, S. (1978). Relationships between size and some life history parameters. *Oecologia (Berlin)*, **37**, 257–72.

Boersma, A. (1974). Opossums in the Hokitika River Catchment. *New Zealand Journal of Forestry Science*, **4**, 64–75.

Bolton, B. L. & Latz, P. K. (1978). The western hare-wallaby, *Lagorchestes hirsutus* (Gould) (Macropodidae) in the Tanami Desert. *Australian Wildlife Research*, **5**, 285–93.

Bolton, B. L., Newsome, A. E. & Merchant, J. C. (1982). Reproduction in the agile wallaby *Macropus agilis* (Gould) in the tropical lowlands of the Northern Territory: opportunism in a seasonal environment. *Australian Journal of Ecology*, **7**, 261–77.

Boonstra, R. (1980). Infanticide in microtines: importance in natural populations. *Oecologia (Berlin)*, **46**, 262–5.

Boucher, D. H. (1977). On wasting parental investment. *American Naturalist*, **111**, 786–8.

Bowley, E. A. (1938). Delayed fertilisation in *Dromicia*. *Journal of Mammalogy*, **20**, 499.

Bradley, A. J., McDonald, I. R. & Lee, A. K. (1980). Stress and mortality in a small marsupial (*Antechinus stuartii* Macleay). *General and Comparative Endocrinology*, **40**, 188–200.

Braithwaite, R. W. (1974). Behavioural changes associated with the population cycle of *Antechinus stuartii* (Marsupialia). *Australian Journal of Zoology*, **22**, 45–62.

Braithwaite, R. W. (1979). Social dominance and habitat utilization in *Antechinus stuartii* (Marsupialia). *Australian Journal of Zoology*, **27**, 517–28.

Braithwaite, R. W., Cockburn, A. & Lee, A. K. (1978). Resource partitioning by small mammals in lowland heath communities of southeastern Australia. *Australian Journal of Ecology*, **3**, 423–45.

Braithwaite, R. W. & Lee, A. K. (1979). A mammalian example of semelparity. *American Naturalist*, **113**, 151–6.

Briand, F. & Yodzis, P. (1982). The phylogenetic distribution of obligate mutualism: evidence of limiting similarity and global instability. *Oikos*, **39**, 273–4.

Brockie, R. E., Bell, B. D. & White, A. J. (1981). Age structure and mortality of possum *Trichosurus vulpecula* populations from New Zealand. *Zoology Publications from the Victoria University of Wellington*, **74**, 63–83.

Brockway, J. M., McDonald, J. D. & Pullar, J. D. (1963). The energy cost of reproduction in sheep. *Journal of Physiology*, **167**, 318–27.

Bronson, M. T. (1979). Altitudinal variation in the life history of the golden-mantled ground squirrel (*Spermophilus lateralis*). *Ecology*, **60**, 272–9.

Brown, J. L. (1964). The evolution of diversity in avian territorial systems. *Wilson Bulletin*, **76**, 106–9.

Brown, W. L. & Wilson, E. O. (1956). Character displacement. *Systematic Zoology*, **5**, 49–64.

Bruce, H. M. (1960). A block to pregnancy in the mouse caused by proximity of strange males. *Journal of Reproduction and Fertility*, **6**, 451–60.

Buchanan, G. & Fraser, E. A. (1918). The development of the urinogenital system in the Marsupialia with special reference to *Trichosurus vulpecula* Part 1. *Journal of Anatomy*, **53**, 35–96.

Buchmann, O. L. & Guiler, E. R. (1977). Behaviour and ecology of the Tasmanian Devil, *Sarcophilus harrisii*. In *The Biology of Marsupials*, ed. B. Stonehouse & D. Gilmore, pp. 155–68. London: MacMillan.

Bulmer, M. G. & Taylor, P. D. (1980). Dispersal and the sex ratio. *Nature, London*, **284**, 448–9.

Bush, G. L., Case, S. M., Wilson, A. C. & Patton, J. L. (1977). Rapid speciation and chromosomal evolution in mammals. *Proceedings of the National Academy of Sciences, USA*, **74**, 3942–6.

Calaby, J. H. (1960). Observations on the banded anteater, *Myrmecobius f. fasciatus* Waterhouse (Marsupialia), with particular reference to its food habits. *Proceedings of the Zoological Society of London*, **135**, 183–207.

Calaby, J. H. & Poole, W. E. (1971). Breeding kangaroos in captivity. *International Zoo Yearbook*, **11**, 5–12.

Calaby, J. H. & Taylor, J. M. (1981). Reproduction in two marsupial-mice, *Antechinus bellus* and *A. bilarni* (Dasyuridae), of tropical Australia. *Journal of Mammalogy*, **62**, 329–41.

Calder, W. A. (1982). The pace of growth: an allometric approach to comparative embryonic and postembryonic growth. *Journal of Zoology, London*, 215–25.

Calow, P. (1979). The cost of reproduction – a physiological approach. *Biological Reviews of the Cambridge Philosophical Society*, **54**, 23–40.

Calow, P. (1982). Homeostasis and fitness. *American Naturalist*, **120**, 416–19.

Calow, P. & Townsend, C. R. (1981). Energetics, ecology and evolution. In *Physiological Ecology: An Evolutionary Approach to Resource Use*, ed. C. R. Townsend & P. Calow, pp. 3–19. Oxford: Blackwell Scientific Publications.

Cannon, J. R., Bakker, H. R., Bradshaw, S. D. & McDonald, I. R. (1976). Gravity as the sole navigational aid to the newborn quokka. *Nature, London*, **259**, 42.

Canny, M. T. (1973). *Phloem Translocation*. Cambridge: Cambridge University Press.

Carpenter, F. L. (1978). Hooks for mammal pollination? *Oecologia (Berlin)*, **35**, 123–32.

Case, T. J. (1978). On the evolution and adaptive significance of postnatal growth rates in terrestrial vertebrates. *Quarterly Review of Biology*, **53**, 243–81.

Cassanova, J. (1958). The dormouse or pygmy possum. *Walkabout*, **24**, 30–1.

Catt, D. C. (1977). The breeding biology of Bennett's wallaby (*Macropus rufogriseus fruticus*) in South Canterbury, New Zealand. *New Zealand Journal of Zoology*, **4**, 401–11.

Caughley, G. & Lawton, J. H. (1981). Plant–herbivore systems. In *Theoretical Ecology: Principles and Applications*, 2nd edn, ed. R. M. May, pp. 132–66. Oxford: Blackwell Scientific Publications.

Charles-Dominique, P. (1974). Ecology and feeding behaviour of five sympatric lorisids in Gabon. In *Prosimian Biology*, ed. R. D. Martin, G. A. Doyle & A. C. Walker, pp. 131–50. Pittsburgh: University of Pittsburgh Press.

Charles-Dominique, P. (1975). Nocturnality and diurnality: an ecological interpretation of these two modes of life by an analysis of the higher vertebrate fauna in tropical forest ecosystems. In *Phylogeny of the Primates: A Multidisciplinary Approach*, ed. W. P. Luckett & F. S. Szalay, pp. 69–88. New York & London: Plenum Press.

Charles-Dominique, P. (1983). Ecology and social adaptations in didelphid marsupials: comparison with eutherians of similar ecology. In *Advances in the Study of Mammalian Behaviour*, ed. J. F. Eisenberg & D. G. Kleiman, pp. 395–422. Special Publication No. 7, The American Society of Mammalogists.

Charles-Dominique, P., Atramentowicz, M., Charles-Dominique, M., Gerard, H., Hladik, A., Hladik, C. M. & Prevost, M. F. (1981). Les mammiferes frugivores arboricoles nocturnes d'une foret guyanaise: inter-relations plant-animaux. *Revue d'Ecologie (La Terre et la Vie)*, **35**, 341–435.

Charlesworth, B. (1980). *Evolution in Age-Structured Populations*. Cambridge: Cambridge University Press.

Charnov, E. L. & Krebs, J. R. (1974). On clutch-size and fitness. *Ibis*, **116**, 217–19.

Cheal, P. D., Lee, A. K. & Barnett, J. L. (1976). Changes in the haematology of *Antechinus stuartii* (Marsupialia), and their association with male mortality. *Australian Journal of Zoology*, **24**, 299–311.

Chippendale, G. M. (1968). The plants grazed by red kangaroos, *Megaleia rufa* (Desmarest), in Central Australia. *Proceedings of the Linnaean Society of New South Wales*, **93**, 98–110.

Christian, J. J. (1950). The adreno–pituitary system and population cycles in mammals. *Journal of Mammalogy*, **31**, 247–59.

Christian, J. J. (1971). Population density and reproductive efficiency. *Biology of Reproduction*, **4**, 248–94.

Christian, J. J. (1978). Neuro-behavioral endocrine regulation of small mammal populations. In *Populations of Small Mammals under Natural Conditions*, ed. D. P. Snyder, pp. 143–58. The Pymatuning Symposia in Ecology, 5. Pittsburgh: University of Pittsburgh.

Cifelli, R. L. (1981). Patterns of evolution among the Artiodactyla and Perissodactyla (Mammalia). *Evolution*, **35**, 433–40.

Clark, A. B. (1978). Sex ratio and local resource competition in a prosimian primate. *Science*, **201**, 163–5.

Clark, M. J. (1967). Pregnancy in the lactating pigmy possum, *Cercartetus concinnus*. *Australian Journal of Zoology*, **15**, 673–83.

Clark, M. J. & Poole, W. E. (1967). The reproductive system and embryonic diapause in the female grey kangaroo, *Macropus giganteus*. *Australian Journal of Zoology*, **15**, 441–59.

Clements, F. E. (1905). *Research Methods in Ecology*. Lincoln, Nebraska: University Publishing Company.

Clifford, H. T. & Drake, W. E. (1980). Pollination and dispersal in eastern Australian heathlands. In *Ecosystems of the World, 9B. Heathlands and Related Shrublands: Analytical Studies*, ed. R. L. Specht, pp. 39–49. Amsterdam: Elsevier.

Close, R. L. (1977). Recurrence of breeding after cessation of suckling in the marsupial *Perameles nasuta*. *Australian Journal of Zoology*, **25**, 641–5.

Clutton-Brock, T. H., Albon, S. & Guinness, F. E. (1981). Parental investment in male and female offspring in polygynous mammals. *Nature, London*, **289**, 487–9.

Clutton-Brock, T. H., Albon, S. D. & Guinness, F. E. (1982). Competition between female relatives in a matrilocal mammal. *Nature, London*, **300**, 178–80.

Clutton-Brock, T. H. & Harvey, P. H. (1978). Mammals, resources and reproductive strategies. *Nature, London*, **273**, 191–5.

Clutton-Brock, T. H. & Harvey, P. H. (1979). Comparison and adaptation. *Proceedings of the Royal Society, London, B*, **205**, 547–65.

Clutton-Brock, T. H. & Harvey, P. H. (1980). Primates, brains and ecology. *Journal of Zoology, London*, **190**, 309–24.

Clutton-Brock, T. H., Harvey, P. H. & Rudder, B. (1977). Sexual dimorphism, socionomic sex ratio and body weight in primates. *Nature, London*, **269**, 797–9.

Cockburn, A. (1978). The distribution of *Pseudomys shortridgei* (Muridae: Rodentia) and its relevance to that of other heathland *Pseudomys*. *Australian Wildlife Research*, **5**, 213–19.

Cockburn, A. (1979). Conservation of the heath rat, *Pseudomys shortridgei*, in the Grampians area. *Victoria's Resources*, **21**, 9–11.

Cockburn, A. (1981a). Population regulation and dispersion of the smoky mouse, *Pseudomys fumeus*. I. Dietary determinants of microhabitat preference. *Australian Journal of Ecology*, **6**, 231–54.

Cockburn, A. (1981b). Diet and habitat preference of the silky desert mouse, *Pseudomys apodemoides* (Rodentia). *Australian Wildlife Research*, **8**, 475–97.

Cockburn, A., Braithwaite, R. W. & Lee, A. K. (1981). The response of the heath rat, *Pseudomys shortridgei*, to pyric succession: a temporally dynamic life-history strategy. *Journal of Animal Ecology*, **50**, 649–66.

Cockburn, A., Lee, A. K. & Martin, R. W. (1983). Macrogeographic variation in litter size in *Antechinus* (Marsupialia: Dasyuridae). *Evolution*, **37**, 86–95.

Cockburn, A. & Lidicker, W. Z. (1983). Microhabitat heterogeneity and population ecology of an herbivorous rodent, *Microtus californicus*. *Oecologia (Berlin)*, **59**, 167–77.

Cockburn, A., Scott, M. P. & Dickman, C. R. (1984a). Mammalian sex ratios, dispersal and intrasexual kin competition. *Oecologia (Berlin)*, in press.

Cockburn, A., Scott, M. P. & Scotts, D. J. (1984b). Inbreeding avoidance and sex-biased natal dispersal in mammals. *Animal Behaviour*, in press.

Cody, M. L. (1966). A general theory of clutch-size. *Evolution*, **20**, 174–84.

Coe, H. & Isaac, F. (1965). Pollination of the baobab (*Adansonia digitata* L.) by the lesser bush baby (*Galago crassicaudatus*). *East African Wildlife Journal*, **3**, 123–4.

Cole, L. C. (1954). The population consequences of life history phenomena. *Quarterly Review of Biology*, **29**, 103–37.

Comins, H. N., Hamilton, W. D. & May, R. M. (1980). Evolutionarily stable dispersal strategies. *Journal of Theoretical Biology*, **82**, 205–30.

Conaway, C. H. (1958). Maintenance, reproduction and growth of the least shrew in captivity. *Journal of Mammalogy*, **39**, 507–12.

Connell, J. H. (1961). The influence of interspecific competition and other factors on the distribution of the barnacle *Chthalamus stellatus*. *Ecology*, **42**, 710–23.

Connell, J. H. (1980). Diversity and the coevolution of competitors, or the ghost of competition past. *Oikos*, **35**, 131–8.

Connor, E. F. & McCoy, E. D. (1979). The statistics and biology of the species–area relationship. *American Naturalist*, **113**, 791–833.

Connor, E. F. & Simberloff, D. (1979). The assembly of species communities: chance or competition? *Ecology*, **60**, 1132–40.

Cook, R. E. (1979). Asexual reproduction: a further consideration. *American Naturalist*, **113**, 769–72.

Copley, P. B. & Robinson, A. C. (1983). Studies on the yellow-footed rock-wallaby, *Petrogale xanthopus* Gray (Marsupialia:Macropodidae). II. Diet. *Australian Wildlife Research*, **10**, 63–76.

Corbett, L. K. (1975). Geographic distribution and habitat of the marsupial mole, *Notoryctes typhlops*. *Australian Mammalogy*, **1**, 375–8.

Croft, D. B. (1981a). Behaviour of red kangaroos, *Macropus rufus* (Desmarest, 1822), in north-western New South Wales. *Australian Mammalogy*, **4**, 5–58.

Croft, D. B. (1981b). Society behaviour of the euro, *Macropus robustus* (Gould), in the Australian arid zone. *Australian Wildlife Research*, **8**, 13–49.

Croft, D. B. (1982). Some observations on the behaviour of the antilopine wallaroo, *Macropus antilopinus* (Marsupialia: Macropodinae). *Australian Mammalogy*, **5**, 5–13.

Culvenor, C. C. J. (1970). Toxic plants, a re-evaluation. *Search*, **1**, 103–10.

Cuttle, P. (1982a). Life history strategy of the dasyurid marsupial, *Phascogale tapoatafa*. In *Carnivorous Marsupials*, ed. M. Archer, pp. 13–22. Sydney: Royal Zoological Society of New South Wales.

Cuttle, P. (1982b). A preliminary report on aspects of the behaviour of the dasyurid marsupial *Phascogale tapoatafa*. In *Carnivorous Marsupials*, ed. M. Archer, pp. 325–32. Sydney: Royal Zoological Society of New South Wales.

Daly, M. (1979). Why don't male mammals lactate? *Journal of Theoretical Biology*, **78**, 325–45.

Davidson, D. W. & Morton, S. R. (1981a). Myrmecochory in some plants (F. Chenopodiaceae) of the Australian arid zone. *Oecologia (Berlin)*, **50**, 357–66.

Davidson, D. W. & Morton, S. R. (1981b). Competition for dispersal in ant-dispersed plants. *Science*, **213**, 1259–61.

Davies, S. J. J. F. (1960). A note on two small mammals of the Darwin area. *Journal of the Proceedings of the Royal Society of Western Australia*, **43**, 63–6.

Davis, D. E. (1945a). The annual cycle of plants, mosquitoes, birds and mammals in two Brazilian forests. *Ecological Monographs*, **15**, 245–95.

Davis, D. E. (1945b). The home range of some Brazilian mammals. *Journal of Mammalogy*, **26**, 119–27.

Dawkins, R. & Carlisle, T. R. (1976). Parental investment, mate desertion and a fallacy. *Nature, London*, **262**, 131–3.

Dawson, T. J., Denny, M. J. S., Russell, E. M. & Ellis, B. (1975). Water usage and diet preferences of free-ranging kangaroos, sheep and feral goats in the Australian arid zone during summer. *Journal of Zoology, London*, **117**, 1–23.

Dawson, T. J. & Ellis, B. A. (1979). Comparison of the diets of yellow-footed rock-wallabies and sympatric herbivores in western New South Wales. *Australian Wildlife Research*, **6**, 245–54.

Delany, M. J. & Happold, D. C. D. (1979). *Ecology of African Mammals*. London & New York: Longman.

Denny, M. J. S. (1982). Review of planigale (Dasyuridae, Marsupialia) ecology. In *Carnivorous Marsupials*, ed. M. Archer, pp. 131–8. Sydney: Royal Zoological Society of New South Wales.

Denny, M. J. S., Gibson, D. & Read, D. (1979). Some results of field observations on planigales. *Bulletin of the Australian Mammal Society*, **5**, 27.

Detling, J. K., Dyer, M. I., Procter-Gregg, C. & Winn, D. T. (1980). Plant–herbivore interactions: examination of potential effects of bison saliva on growth of *Bouteloua gracilis* (H.B.K.) Lag. *Oecologia (Berlin)*, **45**, 26–31.

Diamond, I. T. & Hall, W. C. (1969). Evolution of neocortex. *Science*, **164**, 251–62.

Diamond, J. M. (1975). The island dilemma: lessons of modern biogeographic studies for the design of natural reserves. *Biological Conservation*, **7**, 129–46.

Diamond, J. M. (1978). Niche shifts and the rediscovery of competition. *American Scientist*, **66**, 322–31.

Diamond, J. M. & Gilpin, M. E. (1982). Examination of the "null" model of Connor and Simberloff for species co-occurrences on islands. *Oecologia (Berlin)*, **52**, 64–74.

Diamond, J. M. & May, R. M. (1981). Island biogeography and the design of natural reserves. In *Theoretical Ecology: Principles and Applications*, ed. R. M. May, pp. 228–52. Oxford: Blackwell Scientific Publications.

Dickman, C. R. (1982a). Some observations of the behaviour and nest utilization of free-living *Antechinus stuartii* (Marsupialia: Dasyuridae). *Australian Mammalogy*, **5**, 75–7.

Dickman, C. R. (1982b). Some ecological aspects of seasonal breeding in *Antechinus* (Dasyuridae, Marsupialia). In *Carnivorous Marsupials*, ed. M. Archer, pp. 139–50. Sydney: Royal Zoological Society of New South Wales.

Dickman, C. R. (1984). Competition and coexistence among the small marsupials of Australia and New Guinea. *Acta Zoologica Fennica*, in press.

Dimpel, H. & Calaby, J. H. (1972). Further observations on the mountain pigmy possum (*Burramys parvus*). *Victorian Naturalist*, **89**, 101–6.

Dobson, F. S. (1982). Competition for mates and predominant juvenile male dispersal in mammals. *Animal Behaviour*, **30**, 1183–92.

Dressler, R. L. (1968). Pollination by euglossine bees. *Evolution*, **22**, 202–10.

Dryden, G. L. (1968). Growth and development of *Suncus murinus* in captivity on Guam. *Journal of Mammalogy*, **49**, 51–63.

Dubost, G. (1975). Le comportement du Chevrotain africain, *Hyemoschus aquaticus* Ogilby (Artiodactyla, Ruminantia). *Zeitschrift für Tierpsychologie*, **37**, 403–48.

Duncan, P. E. (1979). The biology of the honey possum (*Tarsipes spenserae*) (Gray, 1842). B. Sc. (Hons.) thesis, School of Environmental and Life Sciences, Murdoch University, Western Australia.

Dunmire, W. W. (1960). An altitudinal survey of reproduction in *Peromyscus maniculatus*. *Ecology*, **41**, 174–82.

Dunnet, G. M. (1956). A live-trapping study of Australian brush-tailed possums, *Trichosurus vulpecula* Kerr (Marsupialia). *CSIRO Wildlife Research*, **1**, 1–18.

Dunnet, G. M. (1964). A field study of local populations of the brush-tailed possum, *Trichosurus vulpecula* in eastern Australia. *Proceedings of the Zoological Society of London*, **142**, 665–95.

Dwyer, P. D. (1977). Notes on *Antechinus* and *Cercartetus* (Marsupialia) in the New Guinea highlands. *Proceedings of the Royal Society of Queensland*, **88**, 69–73.

Dyer, M. I. (1980). Mammalian epidermal growth factor promotes plant growth. *Proceedings of the National Academy of Sciences, USA*, **77**, 4836–7.

Ealey, E. H. M. (1963). The ecological significance of delayed implantation in a population of the hill kangaroo (*Macropus robustus*). In *Delayed Implantation*, ed. A. C. Enders, pp. 33–48. Chicago: University of Chicago Press.

Eberhard, I. D. (1972). Ecology of the koala, *Phascolarctos cinereus* (Goldfuss) on Flinders Chase, Kangaroo Island. Ph. D. thesis, University of Adelaide, South Australia.

Eberhard, I. D. (1978). Ecology of the koala, *Phascolarctos cinereus* (Goldfuss) Marsupialia:

Phascolarctidae in Australia. In *The Ecology of Arboreal Folivores*, ed. G. G. Montgomery, pp. 315–28. Washington, DC: Smithsonian Institute Press.

Edwards, G. P. & Ealey, E. H. M. (1975). Aspects of the ecology of the swamp wallaby, *Wallabia bicolor* (Marsupialia: Macropodidae). *Australian Mammalogy*, 1, 307–17.

Eisenberg, J. F. (1975). Phylogeny, behavior, and ecology in the Mammalia. In *Phylogeny of the Primates: a Multidisciplinary Approach*, ed. W. P. Luckett & F. S. Szalay, pp. 47–68. New York & London: Plenum Press.

Eisenberg, J. F. (1981). *The Mammalian Radiations: An Analysis of Trends in Evolution, Adaptation and Behavior*. London: The Athlone Press.

Eisenberg, J. F. & Hemmer, H. (1972). *Uncia uncia*. *Mammalian Species*, 20, 1–5.

Eisenberg, J. F. & Wilson, D. E. (1981). Relative brain size and demographic strategies in didelphid marsupials. *American Naturalist*, 118, 1–15.

Ellis, B. A., Russell, E. M., Dawson, T. J. & Harrop, C. J. F. (1977). Seasonal changes in diet preferences of free-ranging red kangaroos, euros and sheep in western New South Wales. *Australian Wildlife Research*, 4, 127–44.

Emlen, S. T. & Oring, L. W. (1977). Ecology, sexual selection, and the evolution of mating systems. *Science*, 197, 215–23.

Enders, R. K. (1935). Mammalian life histories from Barro Colorado Island, Panama. *Bulletin of the Museum of Comparative Zoology, Harvard*, 78, 383–502.

Ewer, R. F. (1968). A preliminary survey of the behaviour in captivity of the dasyurid marsupial, *Sminthopsis crassicaudata* (Gould). *Zeitscrift für Tierpsychologie*, 25, 319–65.

Ewer, R. F. (1973). *The Carnivores*. Ithaca, New York: Cornell University Press.

Faegri, K. & van der Pijl, L. (1979). *The Principles of Pollination Ecology*, 3rd edn. London: Pergamon Press.

Fanning, F. D. (1982). Reproduction, growth and development in *Ningaui* sp. (Dasyuridae, Marsupialia) from the Northern Territory. In *Carnivorous Marsupials*, ed. M. Archer, pp. 23–7. Sydney: Royal Zoological Society of New South Wales.

Felsenstein, J. (1981). Scepticism towards Santa Rosalia, or why are there so few kinds of animals? *Evolution*, 35, 124–38.

Fenchel, T. (1974). Intrinsic rate of natural increase: the relationship with body size. *Oecologia (Berlin)*, 14, 317–26.

Fischer, G. M. (1978). Los pequenos mamiferos de Chile. *Universidad de Concepcion, Gayana, Zoologica*, 4, 1–342.

Fisher, R. A. (1930). *The Genetical Theory of Natural Selection*. Oxford: Oxford University Press.

Fleay, D. (1934). The brush-tailed phascogale. The first record of breeding habits. *Victorian Naturalist*, 51, 88–100.

Fleay, D. (1935). Notes on the breeding of Tasmanian devils. *Victorian Naturalist*, 52, 100–5.

Fleay, D. (1940). Breeding of the Tiger-cat. *Victorian Naturalist*, 56, 159–63.

Fleay, D. (1947). *Gliders of the Gum Trees*. Melbourne: Bread And Cheese Club.

Fleay, D. (1949). The yellow-footed marsupial mouse. *Victorian Naturalist*, 65, 273–7.

Fleay, D. (1952). The Tasmanian or marsupial devil – its habits and family life. *Australian Museum Magazine*, 10, 275–80.

Fleay, D. (1960). Breeding the mulgara. *Victorian Naturalist*, 78, 160–7.

Fleay, D. (1962). The northern quoll, *Satanellus hallucatus*. *Victorian Naturalist*, 78, 288–93.

Fleay, D. (1965). Australia's 'needle-in-a-haystack' marsupial. *Victorian Naturalist*, 82, 195–204.

Fleming, M. R. (1980). Thermoregulation and torpor in the sugar glider, *Petaurus breviceps* (Marsupialia: Petauridae). *Australian Journal of Zoology*, 28, 521–34.

Fleming, M. R. (1982). The thermal strategies of three small possums from south-eastern Australia. Ph. D. thesis, Monash University, Clayton, Victoria, Australia.

Fleming, M. W., Harder, J. D. & Wukie, J. J. (1981). Reproductive energetics of the Virginia opossum compared with some eutherians. *Comparative Biochemistry and Physiology*, **70B**, 645–8.

Fleming, T. H. (1972). Aspects of the population dynamics of three species of opossums in the Panama Canal Zone. *Journal of Mammalogy*, **53**, 619–23.

Fleming, T. H. (1973). The reproductive cycles of three species of opossums and other mammals in the Panama Canal Zone. *Journal of Mammalogy*, **54**, 439–55.

Fleming, T. H. (1979). Life-history strategies. In *Ecology of Small Mammals*, ed. D. M. Stoddart, pp. 1–61. London: Chapman & Hall.

Fleming, T. H. & Rauscher, R. J. (1978). On the evolution of litter size in *Peromyscus leucopus*. *Evolution*, **32**, 45–55.

Fletcher, T. P. (1977). Reproduction in the native cat, *Dasyurus viverrinus* (Shaw). B.Sc. (Hons.) thesis, University of Tasmania, Hobart, Tasmania, Australia.

Flynn, T. T. (1923). The yolk-sac and allantoic placenta in *Perameles*. *Quarterly Journal of Microscopical Science*, **67**, 123–82.

Fogel, R. & Trappe, J. M. (1978). Fungus consumption (mycophagy) in small mammals. *Northwest Science*, **52**, 1–31.

Ford, H. A., Paton, D. C. & Forde, N. (1979). Birds as pollinators of Australian plants. *New Zealand Journal of Botany*, **17**, 509–19.

Fox, B. J. (1981). Niche parameters and species richness. *Ecology*, **62**, 1415–25.

Fox, B. J. (1982a). Fire and mammalian secondary succession in an Australian coastal heath. *Ecology*, **61**, 1332–41.

Fox, B. J. (1982b). A review of dasyurid ecology and speculation on the role of limiting similarity in community organization. In *Carnivorous Marsupials*, ed. M. Archer, pp. 97–116. Sydney: Royal Zoological Society of New South Wales.

Fox, B. J. (1982c). Ecological separation and coexistence of *Sminthopsis murina* and *Antechinus stuartii* (Dasyuridae, Marsupialia): a regeneration niche? In *Carnivorous Marsupials*, ed. M. Archer, pp. 187–97. Sydney: Royal Zoological Society of New South Wales.

Fox, B. J. & McKay, G. M. (1981). Small mammal responses to pyric successional changes in eucalypt forest. *Australian Journal of Ecology*, **6**, 29–42.

Fox, B. J. & Whitford, D. (1982). Polyoestry in a predictable coastal environment: reproduction, growth and development in *Sminthopsis murina* (Dasyuridae, Marsupialia). In *Carnivorous Marsupials*, ed. M. Archer, pp. 39–48. Sydney: Royal Zoological Society of New South Wales.

Franklin, I. R. (1980). Evolutionary change in small populations. In *Conservation Biology: An Evolutionary–Ecological Perspective*, ed. M. E. Soule & B. A. Wilcox, pp. 135–49. Sunderland, Massachusetts: Sinauer.

Freeland, W. J. & Janzen, D. H. (1974). Strategies in herbivory by mammals: the role of plant secondary compounds. *American Naturalist*, **108**, 269–89.

Freeland, W. J. & Winter, J. W. (1975). Evolutionary consequences of eating: *Trichosurus vulpecula* (Marsupialia) and the genus *Eucalyptus*. *Journal of Chemical Ecology*, **1**, 439–55.

French, N. R., Stoddart, D. M. & Bobek, B. (1975). Patterns of demography in small mammal populations. In *Small Mammals: Their Productivity and Population Dynamics*, ed. F. B. Golley, K. Petrusewicz & L. Ryszkowski, pp. 73–102. Cambridge: Cambridge University Press.

Frith, H. J. & Sharman, G. B. (1964). Breeding in wild populations of the red kangaroo, *Megaleia rufa*. *CSIRO Wildlife Research*, **9**, 86–114.

Gaines, M. S. & McClenaghan, L. R. (1980). Dispersal in small mammals. *Annual Review of Ecology and Systematics*, **11**, 163–96.

Gall, B. C. (1980). Aspects of the ecology of the koala, *Phascolarctos cinereus* (Goldfuss), in Tucki Tucki Nature Reserve, New South Wales. *Australian Wildlife Research*, **7**, 167–76.

Ganslosser, U. (1982). Sozialstruktur und soziale kommunication bei Marsupialia. *Zoologischer Anzeiger, Jena*, **209**, 294–310.

Gardiner, A. L. (1973). The systematics of the genus *Didelphis* (Marsupialia: Didelphidae) in north and middle America. *Special Publications, The Museum, Texas Technical University*, **4**, 1–81.

Gause, G. F. (1934). *The Struggle for Existence*. Baltimore: Williams & Wilkins.

Gemmell, R. T. (1979). The fine structure of the luteal cells in relation to the concentration of progesterone in the plasma of the lactating bandicoot, *Isoodon macrourus* (Marsupialia: Peramelidae). *Australian Journal of Zoology*, **27**, 501–10.

Gemmell, R. T. (1981). The role of the corpus luteum of lactation in the bandicoot *Isoodon macrourus* (Marsupialia: Peramelidae). *General and Comparative Endocrinology*, **44**, 13–19.

Gemmell, R. T. (1982). Breeding bandicoots in Brisbane (*Isoodon macrourus*; Marsupialia, Peramelidae). *Australian Mammalogy*, **5**, 187–93.

Ghiselin, M. T. (1974). *The Economy of Nature and the Evolution of Sex*. Berkeley: University of California Press.

Giesel, J. T. (1976). Reproductive strategies as adaptations to life in temporally heterogeneous environments. *Annual Reviews of Ecology and Systematics*, **7**, 57–79.

Gilbert, F. S. (1980). The equilibrium theory of island biogeography: fact or fiction? *Journal of Biogeography*, **7**, 209–35.

Gilpin, M. E. & Diamond, J. M. (1982). Factors contributing to non-randomness in species co-occurrences on islands. *Oecologia (Berlin)*, **52**, 75–84.

Gipps, J. H. W., Taitt, M. J., Krebs, C. J. & Dundjerski, Z. (1980). Male aggression and the population dynamics of the vole, *Microtus townsendii*. *Canadian Journal of Zoology*, **59**, 147–57.

Godfrey, G. K. (1969). Reproduction in a laboratory colony of the marsupial mouse *Sminthopsis larapinta* (Marsupialia: Dasyuridae). *Australian Journal of Zoology*, **17**, 637–54.

Godfrey, G. K. & Crowcroft, P. (1971). Breeding the fat-tailed marsupial mouse *Sminthopsis crassicaudata* in captivity. *International Zoo Yearbook*, **11**, 33–8.

Godsell, J. (1982). The population ecology of the eastern quoll *Dasyurus viverrinus* (Dasyuridae, Marsupialia), in southern Tasmania. In *Carnivorous Marsupials*, ed. M. Archer, pp. 199–207. Sydney: Royal Zoological Society of New South Wales.

Gordon, G. (1974). Movements and activity of the short-nosed bandicoot *Isoodon macrourus* (Gould) (Marsupialia). *Mammalia*, **38**, 405–31.

Gordon, G. & Lawrie, B. C. (1977). The rufescent bandicoot, *Echymipera rufescens* (Peters & Doria) on Cape York Peninsula. *Australian Wildlife Research*, **5**, 41–5.

Gosling, L. M. & Petrie, M. (1981). Economics of social organisation. In *Physiological Ecology: An Evolutionary Approach to Resource Use*, ed. C. R. Townsend & P. Calow, pp. 315–45. Oxford: Blackwell Scientific Publications.

Gould, E. & Eisenberg, J. F. (1966). Notes on the biology of the Tenrecidae. *Journal of Mammalogy*, **47**, 660–86.

Gould, S. J. (1966). Allometry and size in ontogeny and phylogeny. *Biological Reviews of the Cambridge Philosophical Society*, **41**, 587–640.

Gould, S. J. (1971). Geometric similarity in allometric growth: a contribution to the problem of scaling in the evolution of size. *American Naturalist*, **105**, 113–36.

Gould, S. J. (1975). Allometry in primates, with emphasis on scaling and the evolution of the brain. *Contributions to Primatology*, **5**, 244–92.

Gould, S. J. (1977). *Ontogeny and Phylogeny*. Cambridge, Massachusetts: Belknap.

Gould, S. J. (1982). Change in developmental timing as a mechanism of macroevolution. In *Evolution and Development*, ed. J. T. Bonner, pp. 333–46. Berlin, Heidelberg & New York: Springer-Verlag.

Gould, S. J. & Lewontin, R. C. (1979). The spandrels of San Marco and the Panglossian paradigm: a critique of the adaptationist programme. *Proceedings of the Royal Society of London, B,* **205**, 581–98.

Grand, T. I. (1983). Body weight in relationship to tissue composition, segmental distribution of mass, and motor function. III. The Didelphidae of French Guyana. *Australian Journal of Zoology,* **31**, 299–312.

Grant, P. R. & Abbott, I. (1980). Interspecific competition, island biogeography, and null hypotheses. *Evolution,* **34**, 332–41.

Grant, P. R. & Grant, B. R. (1980). The breeding and feeding characteristics of Darwin's finches on Isla Genovesa, Galapagos. *Ecological Monographs,* **50**, 381–410.

Green, B., Newgrain, K. & Merchant, J. (1980). Changes in milk composition in the tammar wallaby. *Australian Journal of Biological Sciences,* **33**, 35–42.

Green, R. H. (1967). Notes on the devil (*Sarcophilus harrisii*) and the quoll (*Dasyurus viverrinus*) in northeastern Tasmania. *Records of the Queen Victoria Museum,* **27**, 1–13.

Green, R. H. (1972). The murids and small dasyurids in Tasmania. *Records of the Queen Victoria Museum,* **46**, 1–34.

Greenwood, P. J. (1980). Mating systems, philopatry and dispersal in birds and mammals. *Animal Behaviour,* **28**, 1140–62.

Griffiths, M. & Barker, R. (1966). The plants eaten by sheep and by kangaroos grazing together in a paddock in south-western Queensland. *CSIRO Wildlife Research,* **11**, 145–67.

Griffiths, M., Barker, R. & McLean, L. (1974). Further observations on the plants eaten by kangaroos and sheep grazing together in a paddock in south-western Queensland. *Australian Wildlife Research,* **1**, 27–43.

Griffiths, M., McIntosh, D. L. & Leckie, R. M. C. (1972). The mammary glands of the red kangaroo with observations on the fatty acid components of the milk triglycerides. *Journal of Zoology, London,* **166**, 265–75.

Grossman, G. D. (1982). Dynamics and organisation of a rocky intertidal fish assemblage: the persistence and resilience of taxocene structure. *American Naturalist,* **119**, 611–37.

Guiler, E. R. (1957). Longevity in the wild potoroo, *Potorous tridactylus* (Kerr). *Australian Journal of Science,* **20**, 26.

Guiler, E. R. (1958). Observations on a population of small marsupials in Tasmania. *Journal of Mammalogy,* **39**, 44–58.

Guiler, E. R. (1970a). Observations on the Tasmanian Devil, *Sarcophilus harrisii* (Marsupialia: Dasyuridae). I. Numbers, home range, movements and food in two populations. *Australian Journal of Zoology,* **18**, 49–62.

Guiler, E. R. (1970b). Observations on the Tasmanian Devil, *Sarcophilus harrisii* (Marsupialia: Dasyuridae). II. Reproduction, breeding and growth of young. *Australian Journal of Zoology,* **18**, 63–70.

Guiler, E. R. (1971). Food of the potoroo (Marsupialia, Macropodidae). *Journal of Mammalogy,* **52**, 232–4.

Gullan, P. K. & Norris, K. C. (1981). *An investigation of environmentally significant features (Botanical & Zoological) of Mt. Hotham, Victoria.* Ministry of Conservation, Victoria. Environmental Studies Series Report 315.

Haeckel, E. (1866). *Generelle Morphologie der Organismen: Allgemeine Grundzuge der organischen Formen. Wissenschaft, mechanisch begrundet durch die von Charles Darwin reformirte Descendenz-Theorie.* Berlin: Georg Reimer.

Hall, E. R. & Kelson, K. R. (1959). *The Mammals of North America.* New York: Ronald Press.

Hall, S. (1980a). Diel activity of three small mammals coexisting in forest in southern Victoria. *Australian Mammalogy,* **3**, 67–79.

Hall, S. (1980b). The diets of two coexisting species of *Antechinus* (Marsupialia: Dasyuridae). *Australian Wildlife Research,* **7**, 365–78.

Hall, S. & Lee, A. K. (1982). Habitat use by two species of *Antechinus* and *Rattus fuscipes* in tall open forest in southern Victoria. In *Carnivorous Marsupials*, ed. M. Archer, pp. 209–20. Sydney: Royal Zoological Society of New South Wales.

Hamilton, W. D. (1967). Extraordinary sex ratios. *Science*, **156**, 477–8.

Hanwell, A. & Peaker, M. (1977). Physiological effects of lactation on the mother. *Symposium of the Zoological Society of London*, **41**, 297–312.

Happold, M. (1972). Maternal and juvenile behaviour in the marsupial jerboa *Antechinomys spenceri* (Dasyuridae). *Australian Mammalogy*, **1**, 27–37.

Hardin, G. (1960). The competitive exclusion principle. *Science*, **131**, 1292–7.

Harper, J. L. (1977). The contributions of terrestrial plant studies to the development of the theory of ecology. In *Changing Scenes in the Life Sciences*, ed. C. E. Goulden, pp. 139–57. Pennsylvania: Academy of Natural Sciences, Philadelphia.

Harper, J. L. (1982). After description. In *The Plant Community as a Working Mechanism*, ed. E. I. Newman, pp. 11–25. Oxford: Blackwell Scientific Publications.

Harvey, P. H. (1982). On rethinking allometry. *Journal of Theoretical Biology*, **95**, 37–41.

Harvey, P. H. & Mace, G. M. (1982). Comparisons between taxa and adaptive trends: problems of methodology. In *Current Problems in Sociobiology*, ed. King's College Sociobiology Group, pp. 343–61. Cambridge: Cambridge University Press.

Heidt, G. A., Petersen, M. K. & Kirkland, G. L. (1968). Mating behavior and development of least weasels (*Mustela nivalis*) in captivity. *Journal of Mammalogy*, **49**, 413–19.

Heinsohn, G. E. (1966). Ecology and reproduction of the Tasmanian bandicoots (*Perameles gunnii* and *Isoodon obesulus*). *University of California Publications in Zoology*, **80**, 1–107.

Heinsohn, G. E. (1968). Habitat requirements and reproductive potential of the macropod marsupial *Potorous tridactylus* in Tasmania. *Mammalia*, **32**, 30–43.

Heinsohn, G. E. (1970). World's smallest marsupial: the flat-headed marsupial mouse. *Animals*, **13**, 220–2.

Hendrickson, J. A. (1981). Community-wide character displacement reexamined. *American Naturalist*, **35**, 794–809.

Hennemann, W. W. (1983). Relationship among body mass, metabolic rate and the intrinsic rate of natural increase in mammals. *Oecologia (Berlin)*, **56**, 104–8.

Henry, S. (1984a). Diet and socioecology of gliding possums in southern Victoria. Ph.D. thesis, Monash University, Clayton, Victoria, Australia.

Henry, S. (1984b). Social organisation of the Greater Glider. In *Possums and Gliders*, ed. A. P. Smith & I. D. Hume, in press. Chipping Norton, New South Wales: Surrey Beatty.

Henry, S. & Craig, S. (1984). Diet and home range of the Yellow-bellied Glider in Victoria. In *Possums and Gliders*, ed. A. P. Smith & I. D. Hume, in press. Chipping Norton, New South Wales: Surrey Beatty.

Herrera, C. M. (1982). Grasses, grazers, mutualism, and coevolution: a comment. *Oikos*, **38**, 254–8.

Hill, J. P. (1918). Contributions to the embryology of the Marsupialia. 4. The early development of the Marsupialia with special reference to the native cat (*Dasyurus viverrinus*). *Quarterly Journal of the Microscopical Society*, **56**, 1–134.

Hill, J. P. & Hill, W. C. O. (1955). The growth stages of the pouch young of the native cat (*Dasyurus viverrinus*) together with observations of the anatomy of the new born young. *Transactions of the Zoological Society of London*, **28**, 349–53.

Hill, J. P. & O'Donoghue, C. H. (1913). The reproductive cycle in the marsupial (*Dasyurus viverrinus*). *Quarterly Journal of the Microscopical Society*, **59**, 133–74.

Hindell, M. A., Handasyde, K. A. & Lee, A. K. (1985). Tree species selection by free-ranging koala populations in Victoria. *Australian Wildlife Research*, **12**, in press.

Hinton, H. E. & Dunn, A. M. S. (1967). *Mongooses*. Edinburgh: Oliver & Boyd.

Hladik, C. M. (1979). Diet and ecology of Prosimians. In *The Study of Prosimian Behaviour,* ed. G. A. Doyle & R. D. Martin, pp. 307–57. New York: Academic Press.

Hladik, C. M., Charles-Dominique, P. & Petter, J. J. (1980). Feeding strategies of five nocturnal prosimians in the dry forest of the west coast of Madagascar. In *Nocturnal Malagasy Primates,* ed. P. Charles-Dominique, pp. 41–73. New York: Academic Press.

Hocking, G. J. (1981). The population ecology of the brush-tailed possum, *Trichosurus vulpecula* (Kerr), in Tasmania. M.Sc. thesis, University of Tasmania.

Hofman, M. A. (1982a). Encephalization in mammals in relation to the size of the cerebral cortex. *Brain, Behavior and Evolution,* 20, 84–96.

Hofman, M. A. (1982b). A two-component theory of encephalization in mammals. *Journal of Theoretical Biology,* 99, 571–84.

Hogstedt, G. (1980). Evolution of clutch size in birds: adaptive variation in relation to territory quality. *Science,* 210, 1148–50.

Hoogland, J. L. (1981). Sex ratio and local resource competition. *American Naturalist,* 117, 796–7.

Hopper, S. D. (1980). Bird and mammal pollen vectors in *Banksia* communities at Cheyne Beach, Western Australia. *Australian Journal of Botany,* 28, 61–76.

Hopper, S. D. & Burbidge, A. A. (1982). Feeding behaviour of birds and mammals on flowers of *Banksia grandis* and *Eucalyptus angulosa.* In *Pollination and Evolution,* ed. J. A. Armstrong, J. M. Powell & A. J. Richards, pp. 67–75. Sydney: Royal Botanic Gardens.

Horn, H. S. (1978). Optimal tactics of reproduction and life history. In *Behavioural Ecology: an Evolutionary Approach,* ed. J. R. Krebs & N. B. Davies, pp. 421–9. Oxford: Blackwell Scientific Publications.

How, R. A. (1976). Reproduction, growth and survival of young in the mountain possum, *Trichosurus caninus* (Marsupialia). *Australian Journal of Zoology,* 24, 189–99.

How, R. A. (1978). Population strategies of four species of Australian "possum". In *The Ecology of Arboreal Folivores,* ed. G. G. Montgomery, pp. 305–13. Washington, DC: Smithsonian Institute Press.

How, R. A. (1981). Population parameters of two congeneric possums, *Trichosurus* spp., in north-eastern New South Wales. *Australian Journal of Zoology,* 29, 205–15.

Howe, J. G., Grant, W. E. & Folse, L. J. (1982). Effects of grazing by *Sigmodon hispidus* on the regrowth of annual rye grass (*Lolium perenne*). *Journal of Mammalogy,* 63, 176–9.

Hrdy, S. B. (1979). Infanticide among animals: a review, classification, and examination of the implications for the reproductive strategies of females. *Ethology and Sociobiology,* 1, 13–40.

Hughes, R. L. (1962). Reproduction in the macropod marsupial *Potorous tridactylus* (Kerr). *Australian Journal of Zoology,* 10, 193–224.

Hughes, R. L. (1974). Morphological studies on implantation in marsupials. *Journal of Reproduction and Fertility,* 39, 173–86.

Hughes, R. L. (1982). Reproduction in the Tasmanian devil *Sarcophilus harrisii* (Dasyuridae, Marsupialia). In *Carnivorous Marsupials,* ed. M. Archer, pp. 49–63. Sydney: Royal Zoological Society of New South Wales.

Hughes, R. L., Thomson, J. A. & Owen, W. H. (1965). Reproduction in natural populations of the Australian ringtail possum, *Pseudocheirus peregrinus* (Marsupialia: Phalangeridae) in Victoria. *Australian Journal of Zoology,* 13, 383–406.

Hulbert, A. J. (1980). The evolution of energy metabolism in mammals. In *Comparative Physiology: Primitive Mammals,* ed. K. Schmidt-Nielsen, L. Bolis & C. R. Tracy, pp. 129–47. Cambridge: Cambridge University Press.

Hume, I. D. (1978). Evolution of the Macropodidae digestive system. *Australian Mammalogy,* 2, 37–42.

248   *References*

Hume, I. D. (1982). *Digestive Physiology and Nutrition of Marsupials.* Cambridge: Cambridge University Press.

Hunsaker, D. (Ed.) (1977). *The Biology of Marsupials.* New York: Academic Press.

Hurlbert, S. H. (1981). A gentle depilation of the niche: Dicean resource sets in resource hyperspace. *Evolutionary Theory*, **5**, 177–84.

Hutchinson, G. E. (1957). Concluding remarks. *Cold Spring Harbor Symposia On Quantitative Biology*, **22**, 415–27.

Hutchinson, G. E. (1959). Homage to Santa Rosalia, or why are there so many kinds of animals. *American Naturalist*, **93**, 149–59.

Hutchinson, G. E. (1975). Variations on a theme by Robert MacArthur. In *Ecology and Evolution of Communities*, ed. M. L. Cody & J. M. Diamond, pp. 492–521. Cambridge, Massachusetts: The Belknap Press.

Hutchinson, G. E. (1978). *An Introduction to Population Ecology.* New Haven, Connecticut: Yale University Press.

Hutson, G. D. (1976). Grooming behaviour and birth in the dasyurid marsupial *Dasyuroides byrnei. Australian Journal of Zoology*, **24**, 277–82.

Hutson, G. D. (1982). An analysis of offensive and defensive threat displays in *Dasyuroides byrnei* (Dasyuridae, Marsupialia). In *Carnivorous Marsupials*, ed. M. Archer, pp. 345–63. Sydney: Royal Zoological Society of New South Wales.

Huxley, J. S. (1932). *Problems of Relative Growth.* London: Mac Veagh.

Innes, D. G. L. & Millar, J. G. (1981). Body weight, litter size, and energetics of reproduction in *Clethrionomys gapperi* and *Microtus pennsylvanicus. Canadian Journal of Zoology*, **59**, 785–9.

Inouye, D. W. (1982). The consequences of herbivory: a mixed blessing for *Jurinea mollis* (Asteraceae). *Oikos*, **39**, 269–72.

Jackson, J. B. C. (1981). Interspecific competition and species' distributions: the ghosts of theories and data past. *American Zoologist*, **21**, 889–901.

Janis, C. (1976). The evolutionary strategy of the Equidae and the origins of rumen and cecal digestion. *Evolution*, **30**, 757–74.

Jannett, F. J. (1980). Social dynamics of the montane vole, *Microtus montanus*, as a paradigm. *Biologist*, **62**, 3–19.

Janson, C. H., Terborgh, J. & Emmons, L. H. (1981). Non-flying mammals as pollinating agents in the Amazonian forest. *Biotropica*, **13**, Supplement, 1–6.

Janzen, D. H. (1971). Seed predation by animals. *Annual Review of Ecology and Systematics*, **2**, 465–92.

Janzen, D. H. (1976). The depression of reptile biomass by large herbivores. *American Naturalist*, **110**, 371–400.

Janzen, D. H. (1980). When is it coevolution? *Evolution*, **34**, 611–12.

Janzen, D. H. & Martin, P. S. (1982). Neotropical anachronisms: the fruits the Gomphotheres ate. *Science*, **215**, 19–27.

Jarman, P. J. (1974). The social organisation of antelope in relation to their ecology. *Behaviour*, **48**, 215–67.

Jerison, H. J. (1973). *Evolution of the Brain and Intelligence.* New York and London: Academic Press.

Johnson, C. N. (1983). Variations in group size and composition in red and western grey kangaroos, *Macropus rufus* (Desmarest) and *M. fuliginosus* (Desmarest). *Australian Wildlife Research*, **10**, 25–31.

Johnson, C. N. & Bayliss, P. G. (1983). Habitat selection by sex, age and reproductive class in the red kangaroo, *Macropus rufus*, in western New South Wales. *Australian Wildlife Research*, **8**, 465–74.

Johnson, C. N. & Johnson, K. A. (1983). Behaviour of the bilby, *Macrotis lagotis* (Reid), (Marsupialia: Thylacomyidae) in captivity. *Australian Wildlife Research*, **10**, 77–87.

Johnson, J. I., Kirsch, J. A. W. & Switzer, R. C. (1982). Phylogeny through brain traits: fifteen characters which adumbrate mammalian genealogy. *Brain, Behavior and Evolution*, **20**, 72–83.

Johnson, K. A. (1980a). Diet of the bilby, *Macrotis lagotis*, in the western desert region of central Australia. *Bulletin of the Australian Mammal Society*, **6**, 46–7.

Johnson, K. A. (1980b). Spatial and temporal use of habitat by the red-necked pademelon, *Thylogale thetis* (Marsupialia: Macropodidae). *Australian Wildlife Research*, **7**, 157–66.

Johnson, P. M. (1978). Husbandry of the rufous rat kangaroo *Aepyprymnus rufescens* and the brush-tailed rock wallaby *Petrogale penicillata* in captivity. *International Zoo Yearbook*, **18**, 156–7.

Johnson, P. M. (1979). Reproduction in the plain rock wallaby *Petrogale penicillata inornata* in captivity with age estimation of the pouch young. *Australian Wildlife Research*, **6**, 1–4.

Johnson, P. M. (1980a). Observations of the behaviour of the rufous rat-kangaroo, *Aepyprymnus rufescens* (Gray), in captivity. *Australian Wildlife Research*, **7**, 347–57.

Johnson, P. M. (1980b). Field observations on group compositions in the agile wallaby, *Macropus agilis* (Gould) (Marsupialia: Macropodidae). *Australian Wildlife Research*, **7**, 327–31.

Johnson, P. M. & Bradshaw, I. R. (1977). Rufous rat-kangaroo in Queensland. *Queensland Agricultural Journal*, **103**, 181–3.

Johnson, P. M. & Strahan, R. (1982). A further description of the musky rat-kangaroo, *Hypsiprymnodon moschatus* Ramsay 1876 (Marsupialia, Potoroidae), with notes on its biology. *The Australian Zoologist*, **21**, 27–46.

Jones, F. W. (1923). *The Mammals of South Australia*. Adelaide, South Australia: Government Printer.

Kaczmarski, F. (1966). Bioenergetics of pregnancy and lactation in the Bank vole. *Acta Theriologica*, **11**, 409–17.

Kaufmann, J. H. (1974a). Social ethology of the whiptail wallaby, *Macropus parryi* in northeastern New South Wales, Australia. *Animal Behavior*, **22**, 281–369.

Kaufmann, J. H. (1974b). The ecology and evolution of social organization in the kangaroo family (Macropodidae). *American Zoologist*, **14**, 51–62.

Kaufmann, J. H. (1975). Field observations of the social behaviour of the eastern grey kangaroo, *Macropus giganteus*. *Animal Behaviour*, **23**, 214–21.

Kaufmann, K. W. (1981). Fitting and using growth curves. *Oecologia (Berlin)*, **49**, 293–9.

Kay, R. F. & Hylander, W. L. (1978). The dental structure of mammalian folivores with special reference to primates and phalangeroids (Marsupialia). In *The Ecology of Arboreal Folivores*, ed. G. G. Montgomery, pp. 173–91. Washington, DC: Smithsonian Institution Press.

Keast, A. (1971). Continental drift and the evolution of the biota on southern continents. *Quarterly Review of Biology*, **46**, 335–78.

Keast, A. (1972). Australian mammals: zoogeography and evolution. In *Evolution, Mammals and Southern Continents*, ed. A. Keast, F. C. Erk & B. Glass, pp. 195–246. Albany: State University of New York Press.

Kermack, K. A. & Haldane, J. B. S. (1950). Organic correlation and allometry. *Biometrika*, **37**, 30–41.

Kevan, P. G. (1978). Floral coloration, its colorimetric analysis and significance in anthecology. In *The Pollination of Flowers by Insects*, ed. A. J. Richards, pp. 51–78. London: Academic Press.

Kikkawa, J., Ingram, G. J. & Dwyer, P. D. (1979). The vertebrate fauna of Australian heathlands – an evolutionary perspective. In *Heathlands and Related Shrublands. Descriptive Studies. Ecosystems of the World*, 9A, ed. R. L. Specht, pp. 231–79. Amsterdam: Elsevier.

King, D. R., Oliver, A. J. & Mead, R. J. (1981). *Bettongia* and fluoroacetate: a role for 1080 in fauna management. *Australian Wildlife Research*, **8**, 529–36.

Kinnear, J. E., Cockson, A., Christensen, P. & Main, A. R. (1979). The nutritional biology of the ruminants and ruminant-like mammals – a new approach. *Comparative Biochemistry and Physiology*, **64A**, 357–65.

Kirkpatrick, T. H. (1965). Studies of Macropodidae in Queensland. 3. Reproduction in the grey kangaroo (*Macropus major*) in southern Queensland. *Queensland Journal of Agriculture and Animal Science*, **22**, 319–28.

Kirkpatrick, T. H. & Johnson, P. M. (1969). Studies of Macropodidae in Queensland. 7. Age estimation and reproduction in the agile wallaby (*Wallabia agilis* (Gould)). *Queensland Journal of Agriculture and Animal Science*, **26**, 691–8.

Kirsch, J. A. W. (1977a). The six-percent solution: second thoughts on the adaptedness of the Marsupialia. *American Scientist*, **65**, 276–88.

Kirsch, J. A. W. (1977b). Biological aspects of the marsupial–placental dichotomy: a reply to Lillegraven. *Evolution*, **31**, 898–900.

Kirsch, J. A. W. & Calaby, J. H. (1977). The species of living marsupials: an annotated list. In *The Biology of Marsupials*, ed. B. Stonehouse & D. Gilmore, pp. 9–26. London: Macmillan.

Kirsch, J. A. W. & Waller, P. F. (1979). Notes on the trapping and behavior of the Caenolestidae (Marsupialia). *Journal of Mammalogy*, **60**, 390–5.

Kitchener, D. J. (1973). Notes on home range and movement in two small macropods, the potoroo (*Potorous apicalis*) and the quokka (*Setonix brachyurus*). *Mammalia*, **37**, 231–40.

Kitchener, D. J. (1981). Breeding, diet and habitat preference of *Phascogale calura* (Gould, 1844) (Marsupialia: Dasyuridae) in the southern wheatbelt, Western Australia. *Records of the Western Australian Museum*, **9**, 173–86.

Kitchener, D. J. (1982). Predictors of vertebrate species richness in nature reserves in the Western Australian wheatbelt. *Australian Wildlife Research*, **9**, 1–7.

Kitchener, D. J., Chapman, A., Dell, J., Muir, B. G. & Palmer, M. (1980a). Lizard assemblage and reserve size and structure in the Western Australian wheatbelt – some implications for conservation. *Biological Conservation*, **17**, 25–62.

Kitchener, D. J., Chapman, A., Muir, B. G. & Palmer, M. (1980b). Conservation value for mammals of reserves in the Western Australian wheatbelt. *Biological Conservation*, **18**, 179–207.

Kleiman, D. G. (1977). Monogamy in mammals. *Quarterly Review of Biology*, **52**, 39–69.

Kleiman, D. G. (1981). Correlations among life history characteristics of mammalian species exhibiting two extreme forms of monogamy. In *Natural Selection and Social Behavior*, ed. R. D. Alexander & D. W. Tinkle, pp. 332–44. New York: Chiron Press.

Klomp, H. (1970). The determination of clutch-size in birds: a review. *Ardea*, **58**, 1–124.

Koptur, S. (1979). Facultative mutualism between weedy vetches bearing extrafloral nectaries and weedy ants in California. *American Journal of Botany*, **66**, 1016–20.

Krebs, C. J. & Boonstra, R. (1978). Demography of the spring decline in populations of the vole, *Microtus townsendii*, *Journal of Animal Ecology*, **47**, 1007–15.

Krebs, C. J., Halpin, Z. T. & Smith, J. N. M. (1977). Aggression, testosterone, and the spring decline in populations of the vole *Microtus townsendii*. *Canadian Journal of Zoology*, **55**, 430–7.

Krebs, C. J. & Myers, J. H. (1974). Population cycles in small mammals. *Advances in Ecological Research*, **8**, 267–399.

Krohne, D. T. (1980). Intraspecific litter size variation in *Microtus californicus*. II. Variation between populations. *Evolution*, **34**, 1174–82.

Krohne, D. T. (1981). Intraspecific litter size variation in *Microtus californicus*: variation within populations. *Journal of Mammalogy*, **62**, 29–40.

Labov, J. B. (1981). Pregnancy blocking in rodents: adaptive advantages for females. *American Naturalist*, **118**, 361–71.

Lack, A. (1977). Genets feeding on nectar from *Maranthes polyandra* in Northern Ghana. *East African Wildlife Journal*, **15**, 233–4.

Lack, D. (1947). The significance of clutch size. *Ibis*, **89**, 302–52.

Lack, D. (1954). *The Natural Regulation of Animal Numbers*. Oxford: Clarendon Press.

Lack, D. (1968). *Ecological Adaptations for Breeding in Birds*. London: Methuen.

Laird, A. K. (1966a). Postnatal growth of birds and mammals. *Growth*, **30**, 349–63.

Laird, A. K. (1966b). Dynamics of embryonic growth. *Growth*, **30**, 263–75.

Land Conservation Council (1976). *Report on the Corangamite Study Area*. Melbourne: Land Conservation Council.

Lande, R. (1979). Quantitative genetic analysis of multivariate evolution, applied to brain: body size allometry. *Evolution*, **33**, 402–16.

Law, R. (1980). Ecological determinants in the evolution of life histories. In *Population Dynamics*, ed. R. M. Anderson, B. D. Turner & L. R. Taylor, pp. 81–103. London: Blackwell Scientific Publications.

Lawton, J. H. & Strong, D. R. (1981). Community patterns and competition in folivorous insects. *American Naturalist*, **118**, 317–38.

Lee, A. K., Bradley, A. J. & Braithwaite, R. W. (1977). Corticosteroid levels and male mortality in *Antechinus stuartii*. In *The Biology of Marsupials*, ed. B. Stonehouse & D. Gilmore, pp. 209–20. London: Macmillan.

Lee, A. K. & Cockburn, A. (1984). Spring declines in small mammal populations. *Acta Zoologica Fennica*, **171**, in press.

Lee, A. K. & Nagy, K. A. (1984). Energy and water metabolism in free-living *Antechinus* (Marsupialia:Dasyuridae) during their breeding season. *Australian Journal of Zoology*, in press.

Lee, A. K., Woolley, P. & Braithwaite, R. W. (1982). Life history strategies of dasyurid marsupials. In *Carnivorous Marsupials*, ed. M. Archer, pp. 1–11. Sydney: Royal Zoological Society of New South Wales.

Lemon, M. & Bailey, L. F. (1966). A specific protein difference in the milk from two mammary glands of a red kangaroo. *Australian Journal of Experimental Biology and Medicine*, **44**, 705–8.

Levins, R. & Lewontin, R. (1980). Dialectics and reductionism in ecology. *Synthese*, **43**, 47–78.

Levinton, J. S. (1982). The body size–prey size hypothesis: the adequacy of body size as a vehicle for character displacement. *Ecology*, **63**, 869–72.

Lewontin, R. C. (1979). Fitness, survival and optimality. In *Analysis of Ecological Systems*, ed. D. J. Horn, G. R. Stairs & R. D. Mitchell, pp. 3–21. Columbus: Ohio State University Press.

Lillegraven, J. A. (1969). Latest Cretaceous mammals of the upper part of Edmonton formation of Alberta, Canada, and review of marsupial–placental dichotomy in mammalian evolution. *Paleontological Contributions, University of Kansas*, (Vertebrata 12), 1–122.

Lillegraven, J. A. (1975). Biological considerations of the marsupial–placental dichotomy. *Evolution*, **29**, 707–22.

Lillegraven, J. A. (1979). Reproduction in Mesozoic mammals. In *Mesozoic Mammals: The First Two-thirds of Mammalian History*, ed. J. A. Lillegraven, Z. Kielan-Jaworowska & W. A. Clemens, pp. 259–76. Berkeley: University of California Press.

Lindstedt, S. L. & Calder, W. A. (1981). Body size, physiological time and longevity of homeothermic animals. *Quarterly Review of Biology*, **56**, 1–16.

Linforth, D. J. (1977). The climate of the Otway region. *Proceedings of the Royal Society of Victoria*, **89**, 61–8.

Lithgow, K. A. (1982). Koalas feeding on Monterey Pine. *Victorian Naturalist*, **99**, 259.

Lochmiller, R. L., Whelan, J. B. & Kirkpatrick, R. L. (1982). Energetic cost of lactation in *Microtus pinetorum*. *Journal of Mammalogy*, **63**, 475–81.

Lonneberg, E. & Mjoberg, E. (1916). Results of Dr. E. Mjoberg's Swedish Scientific Expedition to Australia 1910–13. II; Mammals from Queensland. *Kungliga Svenska Vetenskapsakademiens Handlingar, Uppsala and Stockholm*, **52**, 9.

Lord, R. D. (1960). Litter size and latitude in North American mammals. *American Midland Naturalist*, **64**, 488–99.

Low, B. S. (1978). Environmental uncertainty and the parental strategies of marsupials and placentals. *American Naturalist*, **112**, 197–213.

Lumer, C. (1980). Rodent pollination of *Blakea* (Melastomataceae) in a Costa Rican cloud forest. *Brittonia*, **32**, 512–17.

Lyne, A. G. (1964). Observations on the breeding and growth of the marsupial *Perameles nasuta* Geoffroy, with notes on other bandicoots. *Australian Journal of Zoology*, **12**, 322–39.

Lyne, A. G. (1974). Gestation period and birth in the marsupial *Isoodon macrourus*. *Australian Journal of Zoology*, **22**, 303–9.

Lyne, A. G. (1976). Observations on oestrus and the oestrous cycle of the marsupials *Isoodon macrourus* and *Perameles nasuta*. *Australian Journal of Zoology*, **24**, 513–21.

Lyne, A. G. & Hollis, D. E. (1982). Observations on the lateral vaginae and birth canals in the marsupials *Isoodon macrourus* and *Perameles nasuta* (Mammalia). *Journal of Zoology, London*, **198**, 263–77.

Lyne, A. G., Pilton, P. E. & Sharman, G. B. (1959). Oestrous cycle, gestation period and parturition in the marsupial *Trichosurus vulpecula*. *Nature, London*, **183**, 622.

Lyne, A. G. & Verhagen, A. M. W. (1957). Growth of the marsupials *Trichosurus vulpecula* Kerr and *Perameles nasuta* Geoffroy. *Growth*, **21**, 167–95.

MacArthur, R. H. (1972). *Geographical Ecology*. New York: Harper & Row.

MacArthur, R. H. & Levins, R. (1964). Competition, habitat selection, and character displacement in a patchy environment. *Proceedings of the National Academy of Sciences, USA*, **51**, 1207–10.

MacArthur, R. H. & Wilson, E. O. (1967). *The Theory of Island Biogeography*. New Jersey: Princeton University Press.

McCrady, E. (1938). The embryology of the opossum. *American Anatomical Memoirs*, **16**, 1–233.

McDonald, I. R., Lee, A. K., Bradley, A. J. & Than, K. A. (1981). Endocrine changes in dasyurid marsupials with differing mortality patterns. *General and Comparative Endocrinology*, **44**, 292–301.

Mace, G. M. & Eisenberg, J. F. (1982). Competition, niche specialization and the evolution of brain size in the genus *Peromyscus*. *Biological Journal of the Linnaean Society*, **17**, 243–57.

Mace, G. M. & Harvey, P. H. (1983). Energetic constraints on home-range size. *American Naturalist*, **121**, 120–32.

Mace, G. M., Harvey, P. H. & Clutton-Brock, T. H. (1981). Brain size and ecology in small mammals. *Journal of Zoology, London*, **193**, 333–54.

MacFayden, A. (1975). Some thoughts on the behaviour of ecologists. *Journal of Animal Ecology*, **44**, 351–63.

McIntosh, D. L. (1963). Reproduction and growth of the fox in the Canberra district. *CSIRO Wildlife Research*, **8**, 132–41.

McIntosh, R. P. (1980). The background and some current problems of theoretical ecology. *Synthese*, **43**, 195–225.

Mack, G. (1961). Mammals from southwestern Queensland. *Memoirs of the Queensland Museum*, **13**, 213–29.

Mack, R. N. & Thompson, J. N. (1982). Evolution in steppe with few large, hooved mammals. *American Naturalist*, 119, 757–73.

McKay, G. M. (1982). Nomenclature of the gliding possum genera *Petaurus* and *Petauroides* (Marsupialia: Petauridae). *Australian Mammalogy*, 5, 37–9.

McKenzie, N. L. & Archer, M. (1982). *Sminthopsis youngsoni* (Marsupialia: Dasyuridae), the lesser hairy-footed dunnart, a new species from arid Australia. *Australian Mammalogy*, 5, 267–79.

McKey, D. (1975). The ecology of coevolved seed dispersal systems. In *Coevolution of Plants and Animals*, ed. L. E. Gilbert & P. H. Raven, pp. 159–291. Austin: Texas University Press.

McLure, P. A. (1981). Sex-biased litter reduction in food-restricted wood rats (*Neotoma floridana*). *Science*, 211, 1058–60.

McNab, B. K. (1978a). The comparative energetics of neotropical marsupials. *Journal of Comparative Physiology*, 125, 115–28.

McNab, B. K. (1978b). Energetics of arboreal folivores: physiological problems and ecological consequences of feeding on an ubiquitous food supply. In *The Ecology of Arboreal Folivores*, ed. G. G. Montgomery, pp. 153–62. Washington, DC: Smithsonian Institution Press.

McNab, B. K. (1979). The influence of body size on the energetics and distribution of fossorial and burrowing mammals. *Ecology*, 60, 1010–21.

McNab, B. K. (1980a). Food habits, energetics and the population biology of mammals. *American Naturalist*, 116, 106–24.

McNab, B. K. (1980b). Energetics and the limits to a temperature distribution in armadillos. *Journal of Mammalogy*, 61, 606–27.

McNab, B. K. (1982). The physiological ecology of South American mammals. In *Mammalian Biology in South America*, ed. M. A. Mares & H. H. Genoways, pp. 187–207. Pittsburgh: University of Pittsburgh Press.

McNab, B. K. (1983). Energetics, body size, and the limits to endothermy. *Journal of Zoology, London*, 199, 1–29.

McNaughton, S. J. (1979). Grazing as an optimization process: grass–ungulate relationships in the Serengeti. *American Naturalist*, 113, 691–703.

Main, A. R. (1971). Measures of wellbeing in populations of herbivorous macropod marsupials. In *Proceedings of the Advanced Studies Institute on Dynamics of Numbers and Populations*, ed. P. J. den Boer & G. R. Gradwell, pp. 159–73. Wageningen, The Netherlands: Centre for Agricultural Publishing and Documentation.

Main, A. R. & Bakker, H. R. (1981). Adaptation of macropod marsupials to aridity. In *Ecological Biogeography of Australia*, ed. A. Keast, pp. 1490–519. The Hague: Dr. W. Junk.

Main, A. R., Shield, J. W. & Waring, H. (1959). Recent studies in marsupial ecology. In *Biogeography and Ecology in Australia*, ed. A. Keast, R. L. Crocker & C. S. Christian, pp. 315–31. The Hague: Dr. W. Junk.

Maiorana, V. C. (1978). An explanation of ecological and developmental constants. *Nature, London*, 273, 375–7.

Mallory, F. F. & Brooks, R. J. (1980). Infanticide and pregnancy failure: reproductive strategies in the female collared lemming (*Dicrostonyx groenlandicus*). *Biology of Reproduction*, 22, 192–6.

Mangold-Wirz, K. (1966). Cerebralisation und Ontogenesemodus bei Eutheria. *Acta Anatomica*, 63, 449–508.

Mansergh, I. M. & Walsh, N. G. (1983). Observations on the mountain pygmy possum, *Burramys parvus*, on Mt. Higginbotham, Victoria. *Victorian Naturalist*, 100, 106–15.

Margules, C., Higgs, A. J. & Rofe, R. W. (1982). Modern biogeographic theory: are there any lessons for nature reserve design. *Biological Conservation*, 24, 115–28.

Marlow, B. J. (1961). Reproductive behaviour of *Antechinus flavipes* (Waterhouse) (Marsupialia) and the development of the pouch young. *Australian Journal of Zoology*, 9, 203–18.

Marples, T. G. (1973). Studies on the marsupial glider *Schoinobates volans* (Kerr). IV. Feeding biology. *Australian Journal of Zoology*, 21, 213–16.

Marshall, A. J. (1967). Origin of delayed implantation in marsupials. *Nature, London*, 216, 192–3.

Marshall, L. G. (1977). *Lestodelphys halli. Mammalian Species*, 81, 1–3.

Marshall, L. G. (1978a). *Dromiciops australis. Mammalian Species*, 99, 1–4.

Marshall, L. G. (1978b). *Glironia venusta. Mammalian Species*, 107, 1–3.

Marshall, L. G. (1982). Evolution of South American Marsupialia. In *Mammalian Biology in South America*, ed. M. A. Mares & H. H. Genoways, pp. 251–72. Pittsburgh: University of Pittsburgh Press.

Martin, M. M. (1979). Biochemical implications of insect mycophagy. *Biological Reviews of the Cambridge Philosophical Society*, 54, 1–21.

Martin, R. D. (1968). Reproduction and ontogeny in tree shrews (*Tupaia belangeri*) with reference to their general behaviour and taxonomic relationships. *Zeitschrift für Tierpsychologie*, 25, 409–95.

Martin, R. D. (1975). The bearing of reproductive behavior and ontogeny on strepsirhine phylogeny. In *Phylogeny of the Primates: A Multidisciplinary Approach*, ed. W. P. Luckett & F. S. Szalay, pp. 265–97. New York: Plenum.

Martin, R. D. (1981). Relative brain size and basal metabolic rate in terrestrial vertebrates. *Nature, London*, 293, 57–60.

Maser, C., Trappe, J. M. & Nussbaum, R. A. (1978). Fungal–small mammal interrelationships with emphasis on Oregon coniferous forests. *Ecology*, 59, 799–809.

Masters, J. C., Centner, M. R. & Caithness, N. (1982). Sex ratios in galagos revisited. *South African Journal of Science*, 78, 198–202.

Matthews, E. J. (1976). *Insect Ecology*. Brisbane: University of Queensland Press.

Mattingly, D. K. & McClure, P. A. (1982). Energetics of reproduction in large-littered cotton rats (*Sigmodon hispidus*). *Ecology*, 63, 183–95.

May, R. M. (1981). Models for interacting populations. In *Theoretical Ecology: Principles and Applications*, ed. R. M. May, pp. 78–104. Oxford: Blackwell Scientific Publications.

Maynard Smith, J. (1975). *The Theory of Evolution*, 3rd edn. Harmondsworth, Middlesex, United Kingdom: Penguin Books.

Maynard Smith, J. (1977). Parental investment: a prospective analysis. *Animal Behaviour*, 25, 1–9.

Maynard Smith, J. (1978). *The Evolution of Sex*. Cambridge: Cambridge University Press.

Maynard Smith, J. (1982). *Evolution and the Theory of Games*. Cambridge: Cambridge University Press.

Maynes, G. M. (1973a). Reproduction in the parma wallaby, *Macropus parma* Waterhouse. *Australian Journal of Zoology*, 21, 331–51.

Maynes, G. M. (1973b). Aspects of reproduction in the whiptail wallaby, *Macropus parryi*. *Australian Zoologist*, 18, 43–6.

Maynes, G. M. (1974). Occurrence and field recognition of *Macropus parma*. *Australian Zoologist*, 18, 72–87.

Maynes, G. M. (1976). Growth of the Parma wallaby, *Macropus parma* Waterhouse. *Australian Journal of Zoology*, 24, 217–36.

Maynes, G. M. (1977a). Distribution and aspects of the biology of the Parma Wallaby, *Macropus parma*, in New South Wales. *Australian Wildlife Research*, 4, 109–25.

Maynes, G. M. (1977b). Breeding and age structure of the population of *Macropus parma* on Kawau Island, New Zealand. *Australian Journal of Ecology*, 2, 207–14.

Mayr, E. & Provine, W. B. (1980). *The Evolutionary Synthesis: Perspectives on the Unification Of Biology*. Cambridge, Massachusetts: Harvard University Press.

Meijden, van der, E. & van der Waals-Kooi, R. E. (1979). The population ecology of *Senecio jacobaea* in a sand-dune system. I. Reproductive strategy and the biennial habit. *Journal of Ecology*, **67**, 131–53.

Merchant, J. C. (1976). Breeding biology of the agile wallaby *Macropus agilis* (Marsupialia, Macropodidae) in captivity. *Australian Wildlife Research*, **3**, 93–103.

Merchant, J. C. (1979). The effect of pregnancy on the interval between one oestrus and the next in the tammar wallaby, *Macropus eugenii*. *Journal of Reproduction and Fertility*, **56**, 459–63.

Merchant, J. C. & Calaby, J. H. (1981). Reproductive biology of the red-necked wallaby (*Macropus rufogriseus banksianus*) and Bennett's wallaby (*M. r. rufogriseus*) in captivity. *Journal of Zoology, London*, **194**, 203–17.

Merrill, W. & Cowling, E. B. (1966). Role of nitrogen in wood deterioration: amount and distribution of nitrogen in fungi. *Phytopathology*, **56**, 1083–90.

Meyer, J. (1981). A quantitative comparison of the parts of the brains of two Australian marsupials and some eutherian mammals. *Brain, Behavior and Evolution*, **18**, 60–71.

Michener, G. R. (1969). Notes on the breeding and young of the crest-tailed marsupial mouse *Dasycercus cristicauda*. *Journal of Mammalogy*, **50**, 633.

Migula, P. (1969). Bioenergetics of pregnancy and lactation in the European common vole. *Acta Theriologica*, **14**, 167–79.

Miles, M. A., de Souza, A. A. & Povoa, M. M. (1981). Mammal tracking and nest location in Brazilian forest with an improved spool-and-line device. *Journal of Zoology, London*, **195**, 331–47.

Millar, J. S. (1975). Tactics of energy partitioning in breeding *Peromyscus*. *Canadian Journal of Zoology*, **53**, 967–76.

Millar, J. S. (1977). Adaptive features of mammalian reproduction. *Evolution*, **31**, 370–86.

Millar, J. S. (1978). Energetics of reproduction in *Peromyscus leucopus*: the cost of lactation. *Ecology*, **59**, 1055–61.

Millar, J. S. (1979). Energetics of lactation in *Peromyscus maniculatus*. *Canadian Journal of Zoology*, **57**, 1015–19.

Millar, J. S. (1981). Pre-partum reproductive characteristics of eutherian mammals. *Evolution*, **35**, 1149–63.

Milton, K., Windsor, D. M., Morrison, D. W. & Estribi, M. A. (1982). Fruiting phenologies of two neotropical *Ficus* species. *Ecology*, **63**, 752–62.

Moors, P. J. (1974). The foeto-maternal relationship and its significance in marsupial reproduction: a unifying hypothesis. *Australian Mammalogy*, **1**, 263–6.

Moors, P. J. (1975). The urinogenital system and notes on the reproductive biology of the female rufous rat kangaroo *Aepyprymnus rufescens* (Gray) Macropodidae. *Australian Journal of Zoology*, **23**, 355–61.

Morrison, R. G. B. (1975). Emergence of the pygmy *Antechinus*. *Australian Natural History*, **18**, 164–7.

Morton, S. R. (1978a). An ecological study of *Sminthopsis crassicaudata*. III. Reproduction and life history. *Australian Wildlife Research*, **5**, 183–211.

Morton, S. R. (1978b). An ecological study of *Sminthopsis crassicaudata* (Marsupialia: Dasyuridae). II. Behaviour and social organization. *Australian Wildlife Research*, **5**, 163–82.

Morton, S. R. (1979). Diversity of desert-dwelling mammals: a comparison of Australia and North America. *Journal of Mammalogy*, **60**, 253–64.

Morton, S. R. (1982). Dasyurid marsupials of the Australian arid zone: an ecological review. In *Carnivorous Marsupials*, ed. M. Archer, pp. 117–30. Sydney: Royal Zoological Society of New South Wales.

Morton, S. R. & Lee, A. K. (1978). Thermoregulation and metabolism in *Planigale maculata* (Marsupialia: Dasyuridae). *Journal of Thermal Biology*, **3**, 117–20.

Morton, S. R., Recher, H. F., Thompson, S. D. & Braithwaite, R. W. (1982). Comments on the relative advantages of marsupial and eutherian reproduction. *American Naturalist*, 120, 128–34.

Morton, S. R., Wainer, J. W. & Thwaites, T. P. (1980). Distributions and habitats of *Sminthopsis leucopus* and *S. murina* (Marsupialia:Dasyuridae) in south-eastern Australia. *Australian Mammalogy*, 3, 19–30.

Muller, F. (1969). Verhaltnis von Korporentwicklung und Cerebralisation in Ontogenese und Phylogenese der Sauger. *Verhandlungen der Naturforschenden Gesellschaft in Basel*, 80, 1–31.

Muller, F. (1972). Zur stammesgeschichtlichen Veranderung der Eutheria-Ontogenesen. Versuch einer ubersicht aufgrund vergleichent morphologischer studien an Marsupialia und Eutheria. *Revue Suisse de Zoologie*, 79, 1–97 and 501–611.

Muller, F. (1973). Zur stammesgeschichtlichen Veranderung der Eutheria-Ontogenesen. *Revue Suisse de Zoologie*, 79, 1599–685.

Nagy, K. A., Seymour, R. S., Lee, A. K. & Braithwaite, R. W. (1978). Energy and water budgets in free-living *Antechinus stuartii* (Marsupialia:Dasyuridae). *Journal of Mammalogy*, 59, 60–8.

Nelson, J. E. & Stephan, H. (1982). Encephalisation in Australian marsupials. In *Carnivorous Marsupials*, ed. M. Archer, pp. 699–706. Sydney: Royal Zoological Society of New South Wales.

Newsome, A. E. (1964). Anoestrus in the red kangaroo, *Megaleia rufa* (Desmarest). *Australian Journal of Zoology*, 12, 9–17.

Newsome, A. E. (1965). Reproduction in natural populations of the red kangaroo, *Megaleia rufa* (Desmarest), in central Australia. *Australian Journal of Zoology*, 13, 735–59.

Newsome, A. E. (1966). The influence of food on breeding in the red kangaroo in central Australia. *CSIRO Wildlife Research*, 11, 187–96.

Newsome, A. E. (1975). An ecological comparison of the two arid zone kangaroos of Australia, and their anomalous prosperity since the introduction of ruminant stock to their environment. *Quarterly Review of Biology*, 50, 389–424.

Newsome, A. E. (1977). Imbalance in the sex ratio and age structure of the red kangaroo, *Macropus rufus*, in central Australia. In *The Biology of Marsupials*, ed. B. Stonehouse & D. Gilmore, pp. 221–33. London: Macmillan.

Newsome, A. E. & Catling, P. C. (1979). Habitat preferences of mammals inhabiting heathlands of warm temperate coastal, montane and alpine regions of southeastern Australia. In *Heathlands and Related Shrublands: Descriptive Studies*, ed. R. L. Specht, pp. 301–19. Amsterdam: Elsevier.

Newsome, A. E. & Corbett, L. K. (1975). Outbreaks of rodents in central Australia: origins, declines and evolutionary considerations. In *Rodents in Desert Environments*, ed. I. Prakash & P. Ghosh, pp. 117–53. The Hague: Dr. W. Junk.

Nix, H. A. (1976). Environmental control of breeding, post-breeding dispersal and migration of birds in the Australian region. In *Proceedings of the 16th International Ornithological Congress*, ed. H. J. Frith & J. H. Calaby, pp. 272–305. Canberra: Australian Academy of Science.

Noble, J. C. (1975). The effects of emus (*Dromaius novaehollandiae* Latham) on the distribution of the nitre-bush (*Nitraria billardieri* DC). *Journal of Ecology*, 63, 979–84.

O'Connell, M. A. (1979). Ecology of didelphid marsupials from northern Venezuela. In *Vertebrate Ecology in the Northern Neotropics*, ed. J. F. Eisenberg, pp. 73–8. Washington, DC: Smithsonian Institution Press.

Oster, G. F. & Wilson, E. O. (1978). *Caste and Ecology in the Social Insects*. Princeton: Princeton University Press.

Owen, D. F. (1977). Latitudinal gradients in clutch size: an extension of David Lack's theory. In *Evolutionary Ecology*, ed. B. Stonehouse & C. Perrins, pp. 171–9. Baltimore: University Park Press.

Owen, D. F. (1980). How plants may benefit from the animals that eat them. *Oikos*, **35**, 230–5.

Owen, D. F. & Wiegert, R. G. (1976). Do consumers maximise plant fitness. *Oikos*, **27**, 488–92.

Owen, D. F. & Wiegert, R. G. (1981). Mutualism between grasses and grazers: an evolutionary hypothesis. *Oikos*, **36**, 376–8.

Owen, D. F. & Wiegert, R. G. (1982a). Grasses and grazers: is there a mutualism? *Oikos*, **38**, 258–9.

Owen, D. F. & Wiegert, R. G. (1982b). Beating the walnut tree: More on grass/grazer mutualisms. *Oikos*, **39**, 115–16.

Owen, W. H. & Thomson, J. A. (1965). Notes on the comparative ecology of the common brushtail and mountain possums in eastern Australia. *Victorian Naturalist*, **82**, 216–17.

Padykula, H. A. & Taylor, J. M. (1977). Uniqueness of the bandicoot chorioallantoic placenta (Marsupialia: Peramelidae): cytological and evolutionary interpretations. In *Reproduction and Evolution*, ed. J. H. Calaby & C. H. Tyndale-Biscoe, pp. 303–23. Canberra: Australian Academy of Science.

Pahl, L. (1984). Population parameters and diet of the Victorian ringtail possum (*Pseudocheirus peregrinus*). In *Possums and Gliders*, ed. A. P. Smith & I. D. Hume, in press. Chipping Norton, New South Wales: Surrey Beatty.

Park, T. (1954). Experimental studies of interspecific competition: II. Temperature, humidity, and competition in two species of *Tribolium*. *Physiological Zoology*, **27**, 177–238.

Parker, P. (1977). An ecological comparison of marsupial and placental patterns of reproduction. In *The Biology of Marsupials*, ed. B. Stonehouse & D. Gilmore, pp. 273–86. London: Macmillan.

Parker, S. A. (1973). An annotated checklist of the native land mammals of the Northern Territory. *Records of the Australian Museum*, **16**, 1–57.

Parra, R. (1978). Comparison of foregut and hindgut fermentation in herbivores. In *The Ecology of Arboreal Folivores*, ed. G. G. Montgomery, pp. 205–229. Washington, DC: Smithsonian Institution Press.

Paton, D. (1979). Ecology of the New Holland honeyeater. Ph.D. thesis, Monash University, Clayton, Victoria.

Paton, D. C. (1981). The significance of pollen in the diet of the New Holland honeyeater *Phylidonyris novaehollandiae* (Aves: Meliphagidae). *Australian Journal of Zoology*, **29**, 217–24.

Pelikan, J. (1981). Patterns of reproduction in the house mouse. *Symposium of the Zoological Society of London*, **47**, 205–29.

Perrins, C. M. (1965). Population fluctuations and clutch size in the great tit. *Journal of Animal Ecology*, **34**, 601–47.

Perrins, C. M. (1977). The role of predation in the evolution of clutch size. In *Evolutionary Ecology*, ed. B. Stonehouse & C. Perrins, pp. 181–91. Baltimore: University Park Press.

Petelle, M. (1982). More mutualisms between consumers and plants. *Oikos*, **38**, 125–7.

Peters, R. H. (1976). Tautology in evolution and ecology. *American Naturalist*, **110**, 1–12.

Petersen, K. E. & Yates, T. L. (1980). *Condylura cristata*. *Mammalian Species*, **129**, 1–4.

Pianka, E. R. (1981). Competition and niche theory. In *Theoretical Ecology: Principles and Applications*, ed. R. M. May, pp. 168–96. Oxford: Blackwell Scientific Publications.

Pilton, P. E. & Sharman, G. B. (1962). Reproduction in the marsupial *Trichosurus vulpecula*. *Journal of Endocrinology*, **25**, 119–36.

Poole, W. E. (1973). A study of breeding in grey kangaroos, *Macropus giganteus* Shaw and *M. fuliginosus* (Desmarest) in central New South Wales. *Australian Journal of Zoology*, **21**, 183–212.

Poole, W. E. (1975). Reproduction in the two species of grey kangaroos, *Macropus giganteus* Shaw and *M. fuliginosus* (Desmarest). II. Gestation, parturition and pouch life. *Australian Journal of Zoology*, **23**, 333–54.

Poole, W. E. (1976). Breeding biology and current status of the grey kangaroo *Macropus fuliginosus fuliginosus*, of Kangaroo Island, South Australia. *Australian Journal of Zoology*, **24**, 169–87.

Poole, W. E., Carpenter, S. M. & Wood, J. T. (1982). Growth of grey kangaroos and the reliability of age determination from body measurements. I. The eastern grey kangaroo, *Macropus giganteus*. *Australian Wildlife Research*, **9**, 9–20.

Poole, W. E. & Catling, P. C. (1974). Reproduction in the two species of grey kangaroos, *Macropus giganteus* Shaw and *M. fuliginosus* (Desmarest). I. Sexual maturity and oestrus. *Australian Journal of Zoology*, **22**, 277–302.

Portmann, A. (1939). Die Ontogenese der Saugetiere als Evolutionsproblem. *Biomorphosis*, **1**, 109–29.

Portmann, A. (1965). Uber die Evolution der Tragzeit bei Saugertieren. *Revue Suisse de Zoologie*, **72**, 658–66.

Powell, J. A. & Mackie, R. A. (1966). Biological interrelationships of moths and *Yucca whipplei* (Lepidoptera: Gelechiidae, Blastobasidae, Prodoxidae). *University of California Publications in Entomology*, **42**, 1–46.

Pracy, L. T. (1974). Opossums (1). *New Zealand Nature Heritage*, **3**(32), 873–82.

Prance, G. T. (1980). A note on the probable pollination of *Combretum* by *Cebus* monkeys. *Biotropica*, **12**, 239.

Preston, F. W. (1962a). The canonical distribution of commonness and rarity. *Ecology*, **43**, 185–215.

Preston, F. W. (1962b). The canonical distribution of commonness and rarity. II. *Ecology*, **43**, 410–32.

Primack, R. B. & Silander, J. A. (1975). Measuring the relative importance of different pollinators to plants. *Nature, London*, **255**, 143–4.

Pucek, Z. & Lowe, V. P. W. (1975). Age criteria in small mammals. In *Small Mammals: Their Productivity and Population Dynamics*, ed. F. B. Golley, K. Petrusewicz & L. Ryszkowski, pp. 55–72. Cambridge: Cambridge University Press.

Radinsky, L. (1978). Evolution of brain size in carnivores and ungulates. *American Naturalist*, **112**, 815–31.

Radinsky, L. (1981). Brain evolution in extinct South American ungulates. *Brain, Behavior and Evolution*, **18**, 169–87.

Ralls, K. (1976). Mammals in which females are larger than males. *Quarterly Review of Biology*, **51**, 245–76.

Ralls, K. (1977). Sexual dimorphism in mammals: avian models and unanswered questions. *American Naturalist*, **113**, 618–22.

Ralls, K., Barasch, C. & Minkowski, K. (1975). Behavior of captive mouse deer, *Tragulus napu*. *Zeitschrift für Tierpsychologie*, **37**, 356–78.

Ramirez, W. B. (1970). Host specificity of fig wasps (Agaonidae). *Evolution*, **24**, 680–91.

Rand, A. S. (1937). Some original observations on the habits of *Dactylopsila trivirgata* Gray. *American Museum Novitates*, **957**, 1–7.

Randolph, P. A., Randolph, J. C., Mattingly, K. & Foster, M. M. (1977). Energy costs of reproduction in the cotton rat, *Sigmodon hispidus*. *Ecology*, **58**, 31–45.

Rathbun, G. B. (1979). The social structure and ecology of elephant-shrews. *Zeitschrift für Tierpsychologie*, Supplement 20, 1–77.

Read, D. (1982). Observations on the movements of two arid zone planigales (Dasyuridae: Marsupialia). In *Carnivorous Marsupials*, ed. M. Archer, pp. 227–31. Sydney: Royal Zoological Society of New South Wales.

Read, D. G., Fox, B. J. & Whitford, D. (1983). Some notes on breeding in *Sminthopsis* (Marsupialia: Dasyuridae). *Australian Mammalogy*, **6**, 89–92.

Renfree, M. B. (1972). Influence of the embryo on the marsupial uterus. *Nature, London*, **240**, 475–7.

Renfree, M. B. (1977). Feto-placental influences in marsupial gestation. In *Reproduction and*

Evolution, ed. J. H. Calaby & C. H. Tyndale-Biscoe, pp. 325–32. Canberra: Australian Academy of Science.

Renfree, M. B. (1978). Embryonic diapause in mammals – a developmental strategy. In *Dormancy and Developmental Arrest: Experimental Analysis in Plants and Animals*, ed. M. E. Clutter, pp. 1–46. New York: Academic Press.

Renfree, M. B. (1980). Placental function and embryonic development in marsupials. In *Comparative Physiology: Primitive Mammals*, ed. K. Schmidt-Nielsen, L. Bolis & C. R. Taylor, pp. 269–84. Cambridge: Cambridge University Press.

Renfree, M. B. (1981). Embryonic diapause in marsupials. *Journal of Reproduction and Fertility*, Supplement, **29**, 67–78.

Renfree, M. B. (1983). Marsupial reproduction: the choice between placentation and lactation. *Oxford Reviews of Reproductive Biology*, in press.

Renfree, M. B., Holt, A. B., Green, S. W., Carr, J. P. & Cheek, D. B. (1982). Ontogeny of the brain in a marsupial (*Macropus eugenii*) throughout pouch life. I. Brain growth. *Brain, Behavior and Evolution*, **20**, 57–71.

Renfree, M. B. & Tyndale-Biscoe, C. H. (1973). Intrauterine development after diapause in the marsupial, *Macropus eugenii*. *Developmental Biology*, **32**, 28–40.

Rey, J. R., Strong, D. R. & McCoy, E. D. (1982). On overinterpretation of the species-area relationship. *American Naturalist*, **119**, 741–3.

Reynolds, H. C. (1952). Studies on reproduction in the opossum (*Didelphis virginiana*). *University of California Publications in Zoology*, **52**, 233–84.

Richardson, B. J. (1975). r and K selection in kangaroos. *Nature, London*, **255**, 323–4.

Ricklefs, R. E. (1969). An analysis of nestling mortality in birds. *Smithsonian Contributions in Zoology*, **9**, 1–48.

Ricklefs, R. E. (1973). Patterns of growth in birds. II. Growth rate and mode of development. *Ibis*, **115**, 177–201.

Ricklefs, R. E. (1977a). A note on the evolution of clutch size in altricial birds. In *Evolutionary Ecology*, ed. B. Stonehouse & C. Perrins, pp. 193–214. Baltimore: University Park Press.

Ricklefs, R. E. (1977b). On the evolution of reproductive strategies in birds: reproductive effort. *American Naturalist*, **111**, 453–78.

Ricklefs, R. E. (1979). Adaptation, constraint, and compromise in avian development. *Biological Reviews of the Cambridge Philosophical Society*, **54**, 269–90.

Ricklefs, R. E. (1980). Geographical variation in clutch size among passerine birds. Ashmole's hypothesis. *Auk*, **97**, 38–49.

Ride, W. D. L. (1970). *A Guide to the Native Mammals of Australia*. London: Oxford University Press.

Ride, W. D. L. & Tyndale-Biscoe, C. H. (1962). The results of an expedition to Bernier and Dorre Islands, Shark Bay, Western Australia, in July, 1959. Mammals. *Western Australian Fisheries Bulletin*, **2**, 54–97.

Robbins, C. T. & Robbins, B. L. (1979). Fetal and neonatal growth patterns and maternal reproductive effort in ungulates and subungulates. *American Naturalist*, **114**, 101–16.

Robinson, A. C. (1976). Some aspects of the population ecology of the bush rat, *Rattus fuscipes* (Waterhouse). Ph.D. thesis, Monash University, Clayton, Victoria, Australia.

Rose, R. W. (1978). Reproduction and evolution in female Macropodidae. *Australian Mammalogy*, **2**, 65–72.

Rose, R. W. & McCartney, D. J. (1982). Reproduction of the red-bellied pademelon, *Thylogale billardieri* (Marsupialia). *Australian Wildlife Research*, **9**, 27–32.

Rotenberry, J. T. & Wiens, J. A. (1980a). Temporal variation in habitat structure and shrubsteppe bird dynamics. *Oecologia (Berlin)*, **47**, 1–9.

Rotenberry, J. T. & Wiens, J. A. (1980b). Habitat structure, patchiness, and avian communities in North American shrubsteppe vegetation: a multivariate analysis. *Ecology*, **61**, 1228–50.

Roth, V. L. (1979). Can quantum leaps in body size be recognized among mammalian species? *Paleobiology*, **5**, 318–36.

Roth, V. L. (1981). Constancy in the size ratios of sympatric species. *American Naturalist*, **118**, 394–404.

Rothchild, I. (1981). The regulation of the mammalian corpus luteum. *Recent Progress in Hormone Research*, **37**, 183–283.

Roughgarden, J. (1979). *Theory of Population Genetics and Evolutionary Ecology: An Introduction.* New York: Macmillan.

Rourke, J. & Wiens, D. (1977). Convergent floral evolution in South African and Australian Proteaceae and its possible bearing on pollination by non-flying mammals. *Annals of the Missouri Botanical Garden*, **64**, 1–17.

Russell, E. M. (1979). The size and composition of groups in the red kangaroo, *Macropus rufus. Australian Wildlife Research*, **6**, 237–44.

Russell, E. M. (1982a). Patterns of parental care and parental investment in marsupials. *Biological Reviews of the Cambridge Philosophical Society*, **57**, 423–86.

Russell, E. M. (1982b). Parental investment and desertion of young in marsupials. *American Naturalist*, **119**, 744–8.

Russell, E. M. (1984). Social behaviour and social organization of marsupials. *Mammal Review*, in press.

Russell, E. M. & Richardson, B. J. (1971). Some observations on the breeding, age structure, dispersion and habitat of populations of *Macropus robustus* and *Macropus antilopinus* (Marsupialia). *Journal of Zoology, London*, **165**, 131–42.

Russell, R. (1980). *Spotlight on Possums.* St Lucia: University of Queensland Press.

Sacher, G. A. & Staffeldt, E. F. (1974). Relation of gestation time to brain weight for placental mammals: implications for the theory of vertebrate growth. *American Naturalist*, **108**, 593–615.

Sadlier, R. M. F. S. (1965). Reproduction in two species of kangaroo (*Macropus robustus* and *Megaleia rufa*) in the arid Pilbara region of Western Australia. *Proceedings of the Zoological Society of London*, **145**, 239–61.

Sanborn, C. C. (1951). Two new mammals from southern Peru. *Fieldiana: Zoology*, **31**, 473–7.

Sanson, G. D. (1978). The evolution and significance of mastication in the Macropodidae. *Australian Mammalogy*, **2**, 23–8.

Sanson, G. D. (1980). The morphology and occlusion of the molariform cheek teeth in some Macropodinae (Marsupialia: Macropodidae). *Australian Journal of Zoology*, **28**, 341–65.

Sanson, G. D. (1982). Evolution of feeding adaptations in fossil and recent macropodids. In *The Fossil Vertebrate Record of Australia*, ed. P. V. Rich & E. M. Thompson, pp. 489–506. Clayton, Victoria: Monash University Press.

Scarlett, G. & Woolley, P. A. (1980). The honey possum, *Tarsipes spenserae* (Marsupialia:Tarsipedidae): A non-seasonal breeder? *Australian Mammalogy*, **3**, 97–103.

Schemske, D. W. (1982). Ecological correlates of a neotropical mutualism: ant assemblages at *Costus* extrafloral nectaries. *Ecology*, **63**, 932–41.

Schoener, T. W. (1974). Resource partitioning in ecological communities. *Science*, **185**, 27–39.

Schroeder, L. A. (1981). Consumer growth efficiencies: their limits and relationships to ecological energetics. *Journal of Theoretical Biology*, **93**, 805–28.

Schultze-Westrum, T. (1965). Innerartliche verstandigung durch dufle beim gleibeutler *Petaurus breviceps papuanus* Thomas (Marsupialia:Phalangeridae). *Zeitschrift für vergleichende Physiologie*, **50**, 151–220.

Schultze-Westrum, T. G. (1969). Social communication by chemical signals in flying phalangers (*Petaurus breviceps papuanus*). In *Olfaction and Taste*, ed. C. Pfaffman, pp. 268–77. New York: Rockefeller University Press.

Scotts, D. J. (1983). The social organization of *Antechinus stuartii* (Macleay) (Marsupialia,

References 261

Dasyuridae) at Sherbrooke Forest. Victoria. B.Sc. (Hons). thesis, Monash University. Victoria, Australia.

Seebeck, J. H. (1979). Status of the barred bandicoot, *Perameles gunnii*, in Victoria: with a note on husbandry of a captive colony. *Australian Wildlife Research*, 6, 255–64.

Seebeck, J. H. (1981). *Potorous tridactylus* (Kerr) (Marsupialia: Macropodidae): its distribution, status and habitat preferences in Victoria. *Australian Wildlife Research*, 8, 285–306.

Seebeck, J. H. (1982). Breeding, growth and development in *Potorous longipes*. *Bulletin of the Australian Mammal Society*, 7, 48.

Selander, R. K. (1972). Sexual selection and dimorphism in birds. In *Sexual Selection and the Descent of Man*, ed. B. Campbell, pp. 180–320. London: Heinemann.

Selven, H. C. & Stuart, S. (1966). Data dredging procedures in survey analysis. *American Statistician*, 20, 20–3.

Selwood, L. (1980). A timetable of embryonic development of the dasyurid marsupial. *Antechinus stuartii* (Macleay). *Australian Journal of Zoology*, 28, 649–68.

Selwood, L. (1983). Factors influencing pre-natal fertility in the brown marsupial mouse, *Antechinus stuartii*. *Journal of Reproduction and Fertility*, 68, 317–24.

Selye, H. (1936). A syndrome produced by diverse nocuous agents. *Nature, London*, 138, 32.

Sernia, C., Bradley, A. J. & McDonald, I. R. (1980). High affinity binding of adrenocortical and gonadal steroids by plasma proteins of Australian marsupials. *General and Comparative Endocrinology*, 38, 496–503.

Settle, G. A. (1978). The quiddity of Tiger Quolls. *Australian Natural History*, 19, 164–9.

Settle, G. A. & Croft, D. B. (1982a). The development of exploratory behaviour in *Antechinus stuartii* (Dasyuridae, Marsupialia) young in captivity. In *Carnivorous Marsupials*, ed. M. Archer, pp. 383–96. Sydney: Royal Zoological Society of New South Wales.

Settle, G. A. & Croft, D. B. (1982b). Maternal behaviour of *Antechinus stuartii* (Dasyuridae, Marsupialia) in captivity. In *Carnivorous Marsupials*, ed. M. Archer, pp. 365–81. Sydney: Royal Zoological Society of New South Wales.

Shaffer, M. L. (1981). Minimum population sizes for species conservation. *BioScience*, 31, 131–4.

Sharman, G. B. (1955a). Studies on marsupial reproduction 2. The oestrous cycle of *Setonix brachyurus*. *Australian Journal of Zoology*, 3, 44–55.

Sharman, G. B. (1955b). Studies on marsupial reproduction 3. Normal and delayed pregnancy in *Setonix brachyurus*. *Australian Journal of Zoology*, 3, 56–70.

Sharman, G. B. (1961). The embryonic membranes and placentation in five genera of diprotodont marsupials. *Proceedings of the Zoological Society of London*, 137, 197–220.

Sharman, G. B. (1963). Delayed implantation in marsupials. In *Delayed Implantation*, ed. A. C. Enders, pp. 3–14. Chicago: Chicago University Press.

Sharman, G. B. (1965). Marsupials and the evolution of viviparity. In *Viewpoints in Biology*, ed. J. D. Carthy & C. L. Duddington, pp. 1–28. London: Butterworths.

Sharman, G. B. (1973). Adaptations of marsupial young for extra-uterine existence. In *The Mammalian Fetus in Vitro*, ed. C. R. Austin, pp. 67–90. London: Chapman & Hall.

Sharman, G. B. (1982). Karyotypic similarities between *Dromiciops australis* (Microbiotheriidae, Marsupialia) and some Australian marsupials. In *Carnivorous Marsupials*, ed. M. Archer, pp. 711–14. Sydney: Royal Zoological Society of New South Wales.

Sharman, G. B. & Calaby, J. H. (1964). Reproductive behaviour in the red kangaroo, *Megaleia rufa*, in captivity. *CSIRO Wildlife Research*, 9, 58–85.

Sharman, G. B., Calaby, J. H. & Poole, W. E. (1966). Patterns of reproduction in female diprotodont marsupials. *Symposium of the Zoological Society of London*, 15, 205–32.

Sharman, G. B. & Clark, M. J. (1967). Inhibition of ovulation by the corpus luteum in the red kangaroo, *Megaleia rufa*. *Journal of Reproduction and Fertility*, 14, 129–37.

Shaw, G. & Rose, R. W. (1979). Delayed gestation in the potoroo *Potorous tridactylus* (Kerr). *Australian Journal of Zoology*, 27, 901–12.

Sherman, P. W. (1981). Reproductive competition and infanticide in Belding's ground squirrels and other animals. In *Natural Selection and Social Behavior*, ed. R. D. Alexander & D. W. Tinkle, pp. 311–31. New York: Chiron Press.

Shield, J. (1964). A breeding season difference in two populations of the Australian macropod marsupial, *Setonix brachyurus. Journal of Mammalogy*, **45**, 616–25.

Shield, J. (1968). Reproduction of the quokka, *Setonix brachyurus*, in captivity. *Journal of Zoology, London*, **155**, 427–44.

Shield, J. W. & Woolley, P. (1963). Population aspects of delayed birth in the quokka. *Proceedings of the Zoological Society of London*, **141**, 783–9.

Simberloff, D. S. (1980). A succession of paradigms in ecology: essentialism to materialism and probabilism. *Synthese*, **43**, 3–39.

Simberloff, D. S. (1981). Community effects of introduced species. In *Biotic Crises in Ecological and Evolutionary Time*, ed. M. H. Nitecki, pp. 53–81. New York: Academic Press.

Simberloff, D. & Abele, L. G. (1982). Refuge design and island biogeographic theory: effects of fragmentation. *American Naturalist*, **120**, 41–50.

Simberloff, D. & Boecklen, W. (1981). Santa Rosalia reconsidered: size ratios and competition. *Evolution*, **35**, 1206–28.

Simberloff, D., Brown, B. J. & Lowrie, S. (1978). Isopod and insect root borers may benefit Florida mangroves. *Science*, **201**, 630–2.

Sinclair, A. R. E. (1975). The resource limitation of trophic levels in tropical grassland ecosystems. *Journal of Animal Ecology*, **44**, 497–520.

Skutch, A. F. (1967). Adaptive limitation of the reproductive rate of birds. *Ibis,* **109**, 579–99.

Slobodkin, L. B. (1961). *Growth and Regulation in Animal Populations*. New York: Holt, Rinehart & Winston.

Smith, A. P. (1980). The diet and ecology of Leadbeaters possum and the sugar glider. Ph.D. thesis, Monash University, Clayton, Victoria.

Smith, A. P. (1982a). Leadbeaters possum and its management. In *Species at Risk: Research in Australia*, ed. R. H. Groves & W. D. L. Ride, pp. 129–45. Canberra: Australian Academy of Science.

Smith, A. P. (1982b). Diet and feeding strategies of the marsupial sugar glider in temperate Australia. *Journal of Animal Ecology*, **51**, 149–66.

Smith, A. P. (1982c). Is the striped possum (*Dactylopsila trivirgata*; Marsupialia, Petauridae) an arboreal anteater? *Australian Mammalogy*, **5**, 229–34.

Smith, A. P. (1984). Demographic characteristics of Leadbeaters possum, *Gymnobelideus leadbeateri. Australian Wildlife Research*, in press.

Smith, A. P. & Russell, R. P. (1982). Diet of the yellow-bellied glider *Petaurus australis* (Marsupialia:Petauridae) in north Queensland. *Australian Mammalogy*, **5**, 41–5.

Smith, F. E. (1975). Ecosystems and evolution. *Bulletin of the Ecological Society of America*, **56**(4), 2–6.

Smith, G. H. & Taylor, D. J. (1977). Mammary energy metabolism. *Symposium of the Zoological Society of London*, **41**, 95–111.

Smith, M. J. (1973). *Petaurus breviceps. Mammalian Species*, **30**, 1–5.

Smith, M. J. (1979). Observations on growth of *Petaurus breviceps* and *P. norfolcensis* (Petauridae:Marsupialia) in captivity. *Australian Wildlife Research*, **6**, 141–50.

Smith, M. J., Brown, B. K. & Frith, H. J. (1969). Breeding of the brush-tailed possum, *Trichosurus vulpecula* (Kerr) in New South Wales. *CSIRO Wildlife Research*, **14**, 181–93.

Smith, M. J. & How, R. A. (1973). Reproduction in the mountain possum, *Trichosurus caninus* (Ogilby), in captivity. *Australian Journal of Zoology*, **21**, 321–9.

Smith, R. F. C. (1969). Studies of the marsupial glider, *Schoinobates volans* (Kerr). I. Reproduction. *Australian Journal of Zoology*, **17**, 625–36.

Smith, R. J. (1981). *The Tasmanian Tiger – 1980*. Wildlife Division Technical Report 81/1, National Parks and Wildlife Service, Tasmania.

Smith, S. J. (1980). Rethinking allometry. *Journal of Theoretical Biology*, **87**, 97–111.

Smith, S. J. (1981). Interpretation of correlations in intraspecific and interspecific allometry. *Growth*, **45**, 291–7.

Soule, M. E. & Wilcox, B. A. (Eds.) (1980). *Conservation Biology: an Evolutionary–Ecological Perspective*. Sunderland, Massachusetts: Sinauer.

Southwood, T. R. E. (1977). Habitat, the templet for ecological strategies? *Journal of Animal Ecology*, **46**, 337–65.

Southwood, T. R. E. (1980). Ecology – a mixture of pattern and probabilism. *Synthese*, **43**, 111–22.

Specht, R. L. (1973). Structure and functional response of ecosystems in the Mediterranean climate of Australia. In *Mediterranean Type Ecosystems*, ed. F. di Castri & H. A. Mooney, pp. 113–20. London: Chapman & Hall.

Stahl, W. R. (1962). Similarity and dimensional methods in biology. *Science*, **137**, 205–12.

Stark, N. (1972). Nutrient cycling pathways and litter fungi. *Bioscience*, **2**, 355–60.

Statham, H. L. (1982). *Antechinus stuartii* (Dasyuridae: Marsupialia) diet and food availability at Petroi, northeastern New South Wales. In *Carnivorous Marsupials*, ed. M. Archer, pp. 151–63. Sydney: Royal Zoological Society of New South Wales.

Stearns, S. C. (1976). Life history tactics: a review of the ideas. *Quarterly Review of Biology*, **51**, 3–47.

Stearns, S. C. (1977). The evolution of life history traits: a critique of the theory and a review of the data. *Annual Review of Ecology and Systematics*, **8**, 145–71.

Stearns, S. C. (1980). A new view of life-history evolution. *Oikos*, **35**, 266–81.

Stearns, S. C. (1982). The role of development in the evolution of life histories. In *Evolution and Development*, ed. J. T. Bonner, p. 237–58. Berlin: Springer-Verlag.

Stebbins, G. L. (1972). The evolution of the grass family. In *The Biology and Utilization of Grasses*, ed. V. B. Younger & C. M. McKell, pp. 1–17. New York: Academic Press.

Steiner, K. E. (1981). Nectarivory and potential pollination by a neotropical marsupial. *Annals of the Missouri Botanical Garden*, **68**, 505–13.

Stenseth, N. C. (1978). Do grazers maximise individual plant fitness? *Oikos*, **31**, 299–306.

Stephan, H., Nelson, J. E. & Frahm, H. D. (1981). Brain size comparison in Chiroptera. *Zeitschrift für zoologische Systematik und Evolutionforschung*, **19**, 195–222.

Stewart, C. M., Melvin, J. F., Ditchburne, N., Than, S. M. & Zerdoner, E. (1973). The effect of season of growth on chemical composition of cambrial saps in *Eucalyptus regnans* trees. *Oecologia (Berlin)*, **12**, 349–72.

Stodart, E. (1966). Observations on the behaviour of the macropod *Bettongia lesueur* (Quoy and Gaimard) in an enclosure. *CSIRO Wildlife Research*, **11**, 91–9.

Stodart, E. (1977). Breeding and behaviour of Australian bandicoots. In *The Biology of Marsupials*, ed. B. Stonehouse & D. Gilmore, pp. 179–91. London: Macmillan.

Stoddart, D. M. & Braithwaite, R. W. (1979). A strategy for utilization of regenerating heathland habitat by the brown bandicoot (*Isoodon obesulus*; Marsupialia, Peramelidae). *Journal of Animal Ecology*, **48**, 165–79.

Storr, G. M. (1964). Studies in marsupial nutrition. 4. Diet of the quokka, *Setonix brachyurus* (Quoy & Gaimard) on Rottnest Island, Western Australia. *Australian Journal of Biological Sciences*, **17**, 469–81.

Storr, G. M. (1968). Diet of kangaroos (*Megaleia rufa* and *Macropus robustus*) and merino sheep near Port Hedland, Western Australia. *Journal of the Royal Society of Western Australia*, **51**, 25–32.

Strahan, R. (1983). *The Complete Book of Australian Mammals*. Sydney: Angus & Robertson.

Streilein, K. E. (1982a). Behavior, ecology, and distribution of the South American marsupials. In *Mammalian Biology in South America*, ed. M. A. Mares & H. H. Genoways, pp. 231–50. Pittsburgh: University of Pittsburgh Press.

Streilein, K. E. (1982b). The ecology of small mammals in the semiarid Brazilian Caatinga. III. Reproductive biology and population ecology. *Annals of the Carnegie Museum*, **51**, 251–9.

Strong, D. R. (1980). Null hypotheses in ecology. *Synthese*, **43**, 271–86.

Strong, D. R. & Simberloff, D. S. (1981). Straining at gnats and swallowing ratios: character displacement. *Evolution*, **35**, 810–12.

Strong, D. R., Szyska, L. A. & Simberloff, D. S. (1979). Tests of community-wide character displacement against null hypotheses. *Evolution*, **33**, 897–913.

Stuewer, F. W. (1943). Raccoons: their habits and management in Michigan. *Ecological Monographs*, **13**, 203–57.

Suckling, G. C. (1980). The effects of fragmentation and disturbance of forest on mammals in a region of Gippsland, Victoria. Ph.D. thesis, Monash University, Clayton, Victoria, Australia.

Suckling, G. C. (1984). Population ecology of the sugar glider *Petaurus breviceps* in a system of fragmented habitat. *Australian Wildlife Research*, **11**, 49–75.

Sussman, R. W. & Raven, P. H. (1978). Pollination by lemurs and marsupials: an archaic coevolutionary system. *Science*, **200**, 731–6.

Szalay, F. S. (1982). A new appraisal of marsupial phylogeny and classification. In *Carnivorous Marsupials*, ed. M. Archer, pp. 621–40. Sydney: Royal Zoological Society of New South Wales.

Taigen, T. L. (1983). Activity metabolism of anuran amphibians: implications for the origin of endothermy. *American Naturalist*, **121**, 94–109.

Tait, J. F. & Burstein, S. (1964). *In vivo* studies of steroid dynamics in man. In *The Hormones*, vol. 5., ed. G. Pincus, K. V. Thiman & E. B. Astwood, pp. 441–557. New York: Academic Press.

Tamarin, R. H. (1980). Dispersal and population regulation in rodents. In *Biosocial Mechanisms of Population Regulation*, ed. M. N. Cohen, R. S. Malpass & H. G. Klein, pp. 112–33. New Haven, Connecticut: Yale University Press.

Tate, G. H. (1933). A systematic revision of the marsupial genus *Marmosa*. *Bulletin of the American Museum of Natural History*, **66**, 1–250.

Taylor, C. R. (1980). Evolution of mammalian endothermy: a two-step process? In *Comparative Physiology: Primitive Mammals*, ed. K. Schmidt-Nielsen, L. Bolis & C. R. Taylor, pp. 100–11. Cambridge: Cambridge University Press.

Taylor, J. M., Calaby, J. H. & Redhead, T. D. (1982). Breeding in wild populations of the marsupial-mouse *Planigale maculata sinualis* (Dasyuridae, Marsupialia). In *Carnivorous Marsupials*, ed. M. Archer, pp. 83–7. Sydney: Royal Zoological Society of New South Wales.

Taylor, J. M. & Padykula, H. A. (1978). Marsupial trophoblast and mammalian evolution. *Nature, London*, **271**, 588.

Taylor, P. D. (1981). Intra-sex and inter-sex sibling interactions as sex ratio determinants. *Nature, London*, **291**, 64–6.

Taylor, R. J. (1982). Group size in the eastern grey kangaroo, *Macropus giganteus*, and the wallaroo, *Macropus robustus*. *Australian Wildlife Research*, **9**, 229–37.

Temple, S. A. (1977). Plant–animal mutualism: Coevolution with Dodo leads to near extinction of plants. *Science*, **197**, 885–6.

Thomas, O. (1888). Diagnoses of four new species of *Didelphys*. *Annals and Magazine of Natural History* (6), 158–9.

Thomas, O. (1897). Descriptions of new mammals from South America. *Annals and Magazine of Natural History* (7), 265–75.

Thomas, O. (1899). On a new species of *Marmosa*. *Annals and Magazine of Natural History* (7a), 44–5.

Thompson, D. W. (1961). *On Growth and Form*. Cambridge: Cambridge University Press.

Thompson, K. & Uttley, M. G. (1982). Do grasses benefit from grazing? *Oikos*, **39**, 113–15.

Thompson, S. D., MacMillen, R. E., Burke, E. M. & Taylor, C. R. (1980). The energetic cost of bipedal hopping in small mammals. *Nature, London*, **287**, 223–4.

Thomson, J. A. & Owen, W. H. (1964). A field study of the Australian ringtail possum *Pseudocheirus peregrinus* (Marsupialia: Phalangeridae). *Ecological Monographs*, **34**, 27–52.

Tinkle, D. W. (1982). Results of experimental density manipulation in an Arizona lizard community. *Ecology*, **63**, 57–65.

Toro, M. A. (1982). Altruism and sex ratio. *Journal of Theoretical Biology*, **95**, 305–11.

Trappe, J. M. & Maser, C. (1976). Germination of spores of *Glomus macrocarpus* (Endogonaceae) after passage through a rodent digestive tract. *Mycologia*, **68**, 433–6.

Trivers, R. L. (1972). Parental investment and sexual selection. In *Sexual Selection and the Descent of Man, 1871–1971*, ed. B. Campbell, pp. 136–79. Chicago: Aldine.

Trivers, R. L. & Willard, D. E. (1973). Natural selection of parental ability to vary the sex ratio of offspring. *Science*, **179**, 90–2.

Troughton, E. Le G. (1973). *Furred Animals of Australia*. Sydney: Angus & Robertson.

Tuomi, J. (1980). Mammalian reproductive strategies: a generalized relation of litter size to body size. *Oecologia (Berlin)*, **45**, 39–44.

Tuomi, J. & Haukioja, E. (1979a). Predictability of the theory of natural selection: an analysis of the structure of the Darwinian theory. *Savonia*, **3**, 1–8.

Tuomi, J. & Haukioja, E. (1979b). An analysis of natural selection in models of life history theory. *Savonia*, **3**, 9–16.

Turner, B. N. & Iverson, S. L. (1973). The annual cycle of aggression in male *Microtus pennsylvanicus* and its relation to population parameters. *Ecology*, **54**, 967–81.

Turner, V. (1982). Marsupials as pollinators in Australia. In *Pollination and Evolution*, ed. J. A. Armstrong, J. M. Powell & A. J. Richards, pp. 55–66. Sydney: Royal Botanic Gardens.

Turner, V. (1983). Non-flying mammal pollination: an opportunity in Australia. In *Pollination 82*, ed. E. G. Williams, R. B. Knox, J. H. Gilbert & P. Bernhardt, pp. 110–22. Melbourne: Melbourne University Press.

Turner, V. B. (1984a). *Banksia* pollen as a source of protein in the diet of two Australian marsupials, *Cercartetus nanus* and *Tarsipes rostratus*. *Oikos*, **43**, 53–61.

Turner, V. (1984b). *Eucalyptus* pollen in the diet of the feather-tailed glider, *Acrobates pygmaeus* (Marsupialia: Burramyidae). *Australian Wildlife Research*, **11**, 77–81.

Tyndale-Biscoe, C. H. (1955). Observations on the reproduction and ecology of the brush-tailed possum, *Trichosurus vulpecula* Kerr (Marsupialia), in New Zealand. *Australian Journal of Zoology*, **3**, 162–84.

Tyndale-Biscoe, C. H. (1963). Blastocyst transfer in the marsupial, *Setonix brachyurus*. *Journal of Reproduction and Fertility*, **6**, 41–8.

Tyndale-Biscoe, C. H. (1968). Reproduction and post-natal development in the marsupial, *Bettongia lesueur* (Quoy and Gaimard). *Australian Journal of Zoology*, **16**, 577–602.

Tyndale-Biscoe, C. H. (1973). *The Life of Marsupials*. London: Edward Arnold.

Tyndale-Biscoe, C. H., Hearn, J. P. & Renfree, M. B. (1974). Control of reproduction in macropodid marsupials. *Journal of Endocrinology*, **63**, 589–614.

Tyndale-Biscoe, C. H. & MacKenzie, R. B. (1976). Reproduction in *Didelphis marsupialia* and *D. albiventris* in Colombia. *Journal of Mammalogy*, **57**, 249–65.

Tyndale-Biscoe, C. H. & Renfree, M. B. *Reproduction in Marsupials*. Cambridge: Cambridge University Press, in press.

Tyndale-Biscoe, C. H. & Smith, R. F. C. (1969). Studies on the marsupial glider, *Schoinobates volans* (Kerr). II. Population structure and regulatory mechanisms. *Journal of Animal Ecology*, **38**, 637–50.

Vandermeer, J. H. (1972). Niche theory. *Annual Review of Ecology and Systematics*, **3**, 107–32.

Van Dyck, S. (1979). Behaviour in captive individuals of the dasyurid marsupial *Planigale maculata* (Gould, 1851). *Memoirs of the Queensland Museum*, **19**, 413–29.

Van Dyck, S. (1980). The cinnamon antechinus, *Antechinus leo* (Marsupialia: Dasyuridae), a new species from the vine forests of Cape York Peninsula. *Australian Mammalogy*, **3**, 5–17.

Van Dyck, S. (1982a). The relationships of *Antechinus stuartii* and *A. flavipes* (Dasyuridae, Marsupialia) with special reference to Queensland. In *Carnivorous Marsupials*, ed. M. Archer, pp. 723–66. Sydney: Royal Zoological Society of New South Wales.

Van Dyck, S. (1982b). The status and relationships of the Atherton antechinus, *Antechinus godmani* (Marsupialia: Dasyuridae). *Australian Mammalogy*, **5**, 195–210.

Van Soest, P. J. (1982). *Nutritional Ecology of the Ruminant*. Portland, Oregon: Durham & Downey.

Vaughan, T. A. (1969). Reproduction and population densities in a montane small mammal fauna. In *Contributions in Mammalogy*, ed. J. K. Jones, pp. 51–74. Kansas: University of Kansas Museum of Natural History.

Vose, H. M. (1973). Feeding habits of the Western Australian honey possum, *Tarsipes spenserae*. *Journal of Mammalogy*, **54**, 245–7.

Waddington, C. H. (1975). *The Evolution of an Evolutionist*. Ithaca, New York: Cornell University Press.

Wainer, J. W. (1976). Studies of an island population of *Antechinus minimus* (Marsupialia: Dasyuridae). *Australian Zoologist*, **19**, 1–7.

Wakefield, N. A. (1970). Notes on the glider possum, *Petaurus australis*. *Victorian Naturalist*, **87**, 221–36.

Wakefield, N. A. & Warneke, R. M. (1963). Some revisions in *Antechinus* (Marsupialia) – 1. *Victorian Naturalist*, **80**, 194–219.

Wakefield, N. A. & Warneke, R. M. (1967). Some revisions in *Antechinus* (Marsupialia) – 2. *Victorian Naturalist*, **84**, 69–99.

Walker, E. P. (1968). *Mammals of the World*. Baltimore: Johns Hopkins Press.

Walker, K. Z. & Tyndale-Biscoe, C. H. (1978). Immunological aspects of gestation in the tammar wallaby, *Macropus eugenii*. *Australian Journal of Biological Sciences*, **31**, 173–82.

Walker, M. T. & Rose, R. (1981). Prenatal development after diapause in the marsupial *Macropus rufogriseus*. *Australian Journal of Zoology*, **29**, 167–87.

Waring, H. (1956). Marsupial studies in Western Australia. *Australian Journal of Science*, **18**, 66–73.

Warneke, R. M. (1971). Field study of the Australian bush rat, *Rattus fuscipes* (Rodentia: Muridae). *Wildlife Contributions, Fisheries and Wildlife Department, Victoria*, **14**, 1–115.

Waser, P. M. (1980). Small nocturnal carnivores: ecological studies in the Serengeti. *African Journal of Ecology*, **18**, 167–85.

Watts, C. H. S. & Braithwaite, R. W. (1978). The diet of *Rattus lutreolus* and five other rodents in southern Victoria. *Australian Wildlife Research*, **5**, 47–57.

Webb, S. D. (1977). A history of savanna vertebrates in the New World. I. North America. *Annual Review of Ecology and Systematics*, **8**, 355–80.

Wells, R. T. (1978). Field observations of the hairy-nosed wombat, *Lasiorhinus latifrons* (Owen). *Australian Wildlife Research*, **5**, 299–303.

Werren, J. H. & Charnov, E. L. (1978). Facultative sex ratios and population dynamics. *Nature, London*, **272**, 349–50.

Western, D. (1979). Size, life history and ecology in mammals. *African Journal of Ecology*, **17**, 185–204.

Wheelwright, N. T. & Orians, G. H. (1982). Seed dispersal by animals: contrasts with pollen dispersal, problems of terminology, and constraints on coevolution. *American Naturalist,* **119**, 402–13.

Whitford, D., Fanning, F. D. & White, A. W. (1982). Some information on reproduction, growth and development in *Planigale gilesi* (Dasyuridae, Marsupialia). In *Carnivorous Marsupials,* ed. M. Archer, pp. 77–81. Sydney: Royal Zoological Society of New South Wales.

Whittaker, R. H. & Levin, S. A. (Eds.) (1975). *Niche: Theory and Application.* Benchmark Papers in Ecology 3. Stroudsburg, Pennsylvania: Halsted Press.

Whittaker, R. H., Levin, S. A. & Root, R. B. (1973). Niche, habitat and ecotope. *American Naturalist,* **107**, 321–38.

Wiens, D., Renfree, M. B. & Wooller, R. O. (1979). Pollen loads of honey possums (*Tarsipes spenserae*) and nonflying mammal pollination in southwestern Australia. *Annals of the Missouri Botanical Garden,* **66**, 830–8.

Wiens, D. & Rourke, J. P. (1978). Rodent pollination in southern African *Protea* spp. *Nature, London,* **276**, 71–3.

Wiens, D., Rourke, J. P., Casper, B. B., Rickart, E. A., LaPine, T. R. & Peterson, C. J. (1984). Non-flying mammal pollination of southern African proteas: a non-coevolved system. *Annals of the Missouri Botanical Garden,* in press.

Wiens, J. A. (1981). Single-sample surveys of communities: are the revealed patterns real? *American Naturalist,* **117**, 90–8.

Wiens, J. A. & Rotenberry, J. T. (1980). Patterns of morphology and ecology in grassland and shrubsteppe bird populations. *Ecological Monographs,* **50**, 287–308.

Wiens, J. A. & Rotenberry, J. T. (1981). Morphological size ratios and competition in ecological communities. *American Naturalist,* **117**, 592–9.

Williams, G. C. (1966). *Adaptation and Natural Selection.* New Jersey: Princeton University Press.

Williams, G. C. (1975). *Sex and Evolution.* Princeton: Princeton University Press.

Williams, G. C. (1979). The question of adaptive sex ratio in outcrossed vertebrates. *Proceedings of the Royal Society of London, B,* **205**, 567–80.

Williams, R. & Williams, A. (1982). The life cycle of *Antechinus swainsonii* (Dasyuridae, Marsupialia). In *Carnivorous Marsupials,* ed. M. Archer, pp. 89–95. Sydney: Royal Zoological Society of New South Wales.

Williamson, M. (1981). *Island Populations.* Oxford: Oxford University Press.

Wilson, A. C., Bush, G. L., Case, S. M. & King, M. (1975). Social structuring of mammalian populations and rate of chromosomal evolution. *Proceedings of the National Academy of Sciences, USA,* **72**, 6061–5.

Wilson, D. S. (1975). The adequacy of body size as a niche difference. *American Naturalist,* **109**, 769–84.

Wilson, D. S. (1980). *The Natural Selection of Populations and Communities.* Menlo Park, California: Benjamin-Cummings.

Winter, J. W. (1977). The behaviour and social organization of the brush-tail possum (*Trichosurus vulpecula* Kerr). Ph.D. thesis, University of Queensland, Australia.

Wirz, K. (1950). Zur quantitativen Bestimmung der Rangordnung bei Saugetieren. *Acta Anatomica,* **9**, 134–96.

Wittenberger, J. F. (1979). The evolution of mating systems in birds and mammals. In *Handbook of Behavioural Neurobiology, Vol. 3: Social Behavior and Communication,* ed. P. Marler & J. C. Vandenbergh, pp. 271–349. New York: Plenum Press.

Wood, D. H. (1970). An ecological study of *Antechinus stuartii* (Marsupialia) in a southeast Queensland rainforest. *Australian Journal of Zoology,* **18**, 185–207.

Wood, D. H. (1972). The ecology of *Rattus fuscipes* and *Melomys cervinipes* (Rodentia: Muridae) in south-east Queensland rainforest. *Australian Journal of Zoology,* **19**, 371–92.

Woollard, P. (1971). Differential mortality of *Antechinus stuartii* (Macleay): nitrogen balance and somatic changes. *Australian Journal of Zoology*, **19**, 347–53.

Wooller, R. D., Renfree, M. B., Russell, E. M., Dunning, A., Green, S. W. & Duncan, P. (1981). Seasonal changes in a population of the nectar-feeding marsupial *Tarsipes spencerae* (Marsupialia: Tarsipedidae). *Journal of Zoology, London*, **195**, 267–79.

Woolley, P. (1966a). Reproduction in *Antechinus* spp. and other dasyurid marsupials. *Symposium of the Zoological Society of London*, **15**, 281–94.

Woolley, P. (1966b). Reproductive biology of *Antechinus stuartii* Macleay (Marsupialia: Dasyuridae). Ph.D. thesis, Australian National University, Canberra.

Woolley, P. (1971a). Observations on the reproductive biology of the Dibbler, *Antechinus apicalis* (Marsupialia: Dasyuridae). *Journal of the Proceedings of the Royal Society of Western Australia*, **54**, 99–102.

Woolley, P. (1971b). Maintenance and breeding of laboratory colonies of *Dasyuroides byrnei* and *Dasycercus cristicauda*. *International Zoo Yearbook*, **11**, 351–4.

Woolley, P. (1973). Breeding patterns and the breeding and laboratory maintenance of dasyurid marsupials. *Experimental Animals*, **22**, Supplement, 161–72.

Woolley, P. (1974). The pouch of *Planigale subtilissima* and other dasyurid marsupials. *Journal of the Royal Society of Western Australia*, **57**, 11–15.

Woolley, P. (1977). In search of the Dibbler, *Antechinus apicalis* (Marsupialia: Dasyuridae). *Journal of the Royal Society of Western Australia*, **59**, 111–17.

Woolley, P. A. (1981). *Antechinus bellus*, another dasyurid marsupial with a post-mating mortality of males. *Journal of Mammalogy*, **62**, 381–2.

Woolley, P. A. & Ahern, L. D. (1983). Observations on the ecology and reproductive biology of *Sminthopsis leucopus* (Marsupialia: Dasyuridae). *Proceedings of the Royal Society, Victoria*, **95**, 169–80.

Wright, S. J. & Biehl, C. C. (1982). Island biogeographic distributions: testing for random, regular and aggregated patterns of species occurrence. *American Naturalist*, **119**, 345–57.

Wynne-Edwards, V. C. (1962). *Animal Dispersion in Relation to Social Behaviour*. Edinburgh: Oliver & Boyd.

Zar, J. H. (1974). *Biostatistical Analysis*. Englewood Cliffs, New Jersey: Prentice-Hall.

Ziegler, A. C. (1977). Evolution of New Guinea's marsupial fauna in response to a forested environment. In *The Biology of Marsupials*, ed. B. Stonehouse & D. Gilmore, pp. 117–38. London: MacMillan.

Ziegler, A. C. (1981). *Petaurus abidi*, a new species of glider (Marsupialia: Petauridae) from Papua New Guinea. *Australian Mammalogy*, **4**, 81–8.

Ziegler, A. C. (1982). An ecological check-list of New Guinea recent mammals. In *Biogeography and Ecology of New Guinea*, ed. J. L. Gressitt, pp. 863–94. The Hague: Dr. W. Junk.

Marsupial genus and species index

Body weights and lengths for all extant marsupial species are given in Appendix 1. Genus citations are used in cases where the species is not identified in the text, and for monospecific genera.

Acrobates 18, 21–2, 33, 42, 56, 121, 190–1
Aepyprymnus 41, 126–7
Antechinomys 89, 101
Antechinus 13–15, 36, 41, 56, 94, 100, 105, 162–83
  bellus 88
  flavipes 88, 93, 169, 173, 175, 181, 207, 213
  godmani 88
  leo 88, 207
  melanurus 89, 101
  minimus 73, 88, 173–5, 213
  naso 89, 101
  stuartii 13–15, 33, 51, 57, 62, 67, 87–94, 99, 100–1, 103–4, 145–6, 148, 163–83, 205–7, 213–14
  swainsonii 13–15, 54–5, 62, 87–94, 104, 169, 173, 175, 178, 180, 205–6, 213

Bettongia 41, 126
  gaimardii 127
  lesueur 28, 67, 127, 156
  penicillata 28, 34, 62, 127
Burramys 19–20, 22, 36, 41, 71, 121, 123

Caenolestes 12, 41
Caluromys 10, 12, 41, 109–11, 194
  derbianus 107, 109
  philander 54, 62, 106–11, 208
Caluromysiops 10, 41
Cercartetus, 42
  caudatus 36, 67, 121
  concinnus 121
  nanus 19–22, 36, 121

Chaeropus 41
Chironectes 10, 41, 49, 77, 107

Dactylopsila 24–5, 41, 72
Dasycercus 41, 62, 88, 97, 207
Dasykaluta 41
  rosamondae 88
Dasyuroides 41, 62, 89, 99, 101
Dasyurus 41
  geoffroii 88, 207
  maculatus 62, 88, 97
  viverrinus 14–15, 52, 62, 88, 95, 97
Dendrolagus 28, 40
Didelphis 10, 12, 41, 65, 106, 109, 172, 194
  albiventris 106–8
  marsupialis 12, 106–8, 208
  virginiana 51–2, 56, 62, 105–7
Distoechurus 41
Dorcopsis 40
Dorcopsulus 40
Dromiciops 11–12, 41, 105

Echymipera 41

Glironia 10, 41
Gymnobelideus 22–3, 33–4, 41, 72, 124–6, 147–51, 160, 223

Hemibelideus 72
Hypsiprymnodon 28, 36, 40, 62, 77, 126–7, 159

Isoodon 41, 65
  macrourus 56, 62, 112–14
  obesulus 16–17, 112–18, 149

269

Subject index

New Zealand, marsupials feral in 27,
130–1, 135
niche
breadth 204, 210
definition 203
overlap 204, 210
separation 204, 210
null hypotheses, in community ecology
208–13

olfaction 73, 114, 191
organogenesis 51

paedomorphosis 83–5
parental investment 50–68, 76–7, 102,
110–14, 146–7, 151, 160
peramorphosis 83–5
petaurids 19, 22–6, 52–3, 60–1, 63, 67,
71–2, 123–6, 128–32, 147–52, 155, 160
phalangerids 19, 27, 52–3, 60–1, 63, 67,
72, 128–32, 152–5
philopatry 91, 147, 180–2
placentation 52, 113
Plectostrongylus 164
pollination syndrome 186, 188–9, 191–3
polygyny 58, 100, 114, 146–60, 180, 206
polyoestrus 87, 94–6, 98–9, 101–11, 126
potoroines 71, 77, 126–8, 156, 159
pouch life 90, 95, 97, 126–7, 130, 137

r selection 131–2, 140–1
r_m (intrinsic rate of natural increase) 74,
131
radiations, comparisons between 78
rainforest 9, 12, 16, 22, 28, 36, 159, 194–9
reproduction, frequency
burramyids 120–3
dasyurids 87, 90, 94, 98, 100
macropodines 137–8
petaurids 128
phalangerids 130
potoroines 127
reproduction rate 59–68
reproduction, rescindment 55–9, 140
reproductive effort 86–7, 102–3, 163–4
reproductive value 56
resorption *see* reproduction, rescindment

Rodentia 36, 38, 53, 57, 69, 70, 72, 104,
115–19, 169–71, 207–14

seasonality
breeding 87–9, 92–8, 102–3, 105,
112–13, 123, 130, 132–5, 144–5, 176–9
food 32–3, 37, 93, 102–3, 106–8,
113–14, 120, 123–4, 130, 132–42,
176–9, 192, 195–9, 213
seed dispersal 193–9
semelparity 87, 104–5, 162, 175, 181–3
sex allocation 58, 154, 179–83
sexual dimorphism 91, 100, 146–7, 156–7,
206
sexual maturity
bandicoots and bilbies 112–13
burramyids 121, 123
dasyurids 87, 98, 100
didelphids 107
macropodines 135, 137–8
petaurids 123–4, 126, 129–30
phalangerids 67, 129–31
potoroines 127
social organisation 72, 79, 81, 91–2, 100–1,
110–14, 145–61
species diversity 184, 202–5
sperm storage 90
spring declines 163, 170
stress 163–71
succession 101, 103, 110, 114–19, 131,
197–9, 213, 219
symbiosis 184–5
systems ecology 2

territoriality 44, 147, 150–7, 160
trade-offs 46, 64, 172

vision 73, 79

weaning, age at
burramyids 121–2
dasyurids 90, 95, 97–9
macropodines 137–8
petaurids 124, 129
phalangerids 129–30
potoroines 127
wombats 14, 19, 29–30, 52–3, 60–1, 63, 72